# HISTORY of

Charles E. Merrill Publishing Company
*A Bell & Howell Company*
Columbus, Ohio 43216

# MATHEMATICS

**Arthur Gittleman**

California State University
Long Beach

**Merrill Mathematics Series**
Erwin Kleinfeld, *Editor*

Published by
Charles E. Merrill Publishing Co.
*A Bell & Howell Co.*
Columbus, Ohio 43216

Copyright © 1975 by Bell & Howell Company. All rights reserved. No part of this book may be reproduced in any form, electronic or mechanical, including photocopy, recording, or any information storage or retrieval system, without permission in writing from the publisher.

International Standard Book Number: 0–675–08784–8

Library of Congress Catalog Card Number: 74–80376

3 4 5 6 7 8 — 82 81 80 79 78 77 76

Printed in the United States of America

**To
Kaye**

# Preface

Mathematics is appealing. I hope that the reader will find much enjoyment in learning some of the many fascinating mathematical discoveries of the past 5000 years. In addition, the history of mathematics has its own appeal. It not only answers the question of how mathematics was done, but *why* and by whom, when and where.

Mathematics is seen as an aspect of human culture, developing in response both to its environment of social stresses and its heredity of previous mathematics. Anecdotes portray the genius of the great mathematicians and sometimes their eccentricities.

In this book, the history of mathematics is presented chronologically, with specific mathematical examples placed in the context of the general trends of the period. Among the topics treated are: the use of other numeral systems; different methods of arithmetic; the origins of algebra, geometry, trigonometry, analytic geometry, and calculus; and the development of modern mathematics.

I hope that the reader will not only find enjoyment, but also obtain a better appreciation of the usual subjects of high school mathematics and achieve an insight into modern mathematics. Mathematics students should gain a greater perspective of their subject, while teachers and prospective teachers should, in addition, find much that will be of value to them in the classroom.

A background of high school mathematics is assumed. The portions of the latter third of the book which treat more advanced topics are largely self-contained, although a course in calculus would be a desirable prerequisite for these portions.

Problems are found at the end of each chapter. Some will give practice using the methods explained in the book, while others provide supplementary material of interest. Answers to selected problems are given. Suggestions for intended projects are given in Appendix B.

References are listed by author at the end of each chapter. Complete bibliographic information will be found in the Bibliography. Some suggestions for further reading are also appended (A), as is a guide to the pronunciation of names (C).

I gratefully acknowledge my debt to many scholars who have written on various aspects of the history of mathematics. They are too numerous to name here, but are cited in footnotes and references. In addition, I especially thank Mrs. Gene Chapin for beautifully typing the manuscript, and I thank my wife Kaye, who figures in everything I do, for preparing most of the art manuscript and improving the syntax. Professor Erik Hemmingsen of Syracuse University and the staff of Charles E. Merrill Publishing Company have been most helpful. I particularly wish to thank the production editor, Marilyn Schneider, whose expertise has been invaluable.

A.G.

# Contents

**1 Early Mathematics**    1
    1 Egypt    1
    2 Babylonia    9
    3 India    17

**2 Greek Mathematics**    25
    1 The Beginning of Greek Mathematics    26
    2 Crises and the Origin of Deductive Mathematics    31
    3 Greek Number Systems    37
    4 Early Results and Problems of an Independent Mathematics    39
    5 New Methods and Ideas    45
    6 *The Elements* — A Summary    54
    7 The Pinnacle of Greek Geometry    65
    8 The Late Period    75

**3 Mathematics in Asia**    97
    1 The Abacus    97
    2 Chinese Mathematics    100
    3 Indian Mathematics    104
    4 Arabic (Islamic) Mathematics    111

**4 European Mathematics Until 1630**    121
    1 Europe in the Middle Ages (529–1436)    121
    2 The Renaissance    130
    3 Toward Modern Mathematics    139

**5 The Origin and Development of Analytic Geometry and the Calculus**    163
    1 Seventeenth Century Origin    163
    2 Eighteenth Century Development    185

## 6  Mathematics as Free Creation — **205**

   1 A Forerunner — Carl Friedrich Gauss (1777–1855)    206
   2 Advanced Calculus    215
   3 Variety in Geometry    223
   4 Variety in Algebra    227
   5 The Arithmetization of Analysis    233
   6 The Generality of Mathematics in the Twentieth Century    243

## Appendices

   A Suggestions for Further Reading    263
   B Suggestions for Projects    265
   C Guide to the Pronunciation of Names    267
   D Answers to Selected Problems    269

## Bibliography — **277**

## Index — **285**

# History of Mathematics

# Early Mathematics

## Introduction

Mathematics was developed in response to needs of early societies. With growing numbers of people living, working, and even fighting together came the need to solve practical problems of their civilization — problems such as calculating the quantity of materials needed to build a storehouse or the amount of food needed to provision their army. In addition to the practical problems of mathematics were others motivated by religion, including geometric problems arising in the construction of altars and temples. Early mathematics was used in many ways, yet even the earliest mathematical records support the feeling that the solving of mathematical problems was enjoyed for its own sake, too. The three early civilizations whose mathematics we will study in this chapter are Egypt, Babylonia, and India.

## 1 Egypt

Society developed in Egypt along the fertile Nile River. As long ago as 2900 B.C. Egyptian civilization was advanced enough to be able to build one of the wonders of the world, the Great Pyramid of Cheops. No records of any mathematics of that time have been preserved; the main mathematical documents in existence refer to the period of the Middle Kingdom which spanned the years from about 2100 B.C. to 1800 B.C.

It is amazing that any documents at all remain from that period, and, in fact, there are very few extant Egyptian mathematical texts. The only writings which have been preserved are those which were either purposely placed in tombs or were by some accident kept insulated from the elements for thousands of years. The Egyptians wrote on papyrus, a type of paper made from reeds that grew near the water. Being an organic substance, it will soon deteriorate if left to the elements. You can imagine what will happen to the pages of this book in 5000 years, for example, if nothing is done to preserve them. Therefore, it is lucky that some pieces of papyrus have remained as evidence of the extent to which the Egyptians had developed their knowledge of mathematics.

The earliest Egyptian writings, hieroglyphics, are made up of pictorial characters, a picture possibly being the most natural way of representing an object. A house, for instance, might have been represented by a picture of a house. Later the picture became simplified into a conventional sign which was easier to write but looked less like a house. The concept of numbers was also portrayed pictorially by hieroglyphics. The hieroglyphic symbols for numbers are

|  | | = 1
heel bone | ∩ | = 10
snare | ҄ | = 100
lotus flower | ⚶ | = 1000

Symbols also existed for the numbers 100,000 and 1,000,000, but were infrequently used. Other numbers were written using groups of symbols.

532 = ҄҄҄҄҄ ∩∩∩ | |
47 = ∩∩∩∩ | | | | | | |

A main source of information on Egyptian mathematics is a papyrus bought by a Scottish Egyptologist, Rhind, in the nineteenth century, often referred to as the Rhind papyrus. It is also called the Ahmes (A'h-mose) papyrus in honor of the scribe who wrote it. The Ahmes papyrus was copied about 1650 B.C. from an older work of the Middle Kingdom and is a collection of solved problems probably used in a school for scribes. Very few people could write so being a scribe was a revered profession, and this prestige earned the scribe a position close to the people in power. Scribes were afforded such respect that their profession was honored by a famous statue of a scribe dated about 2500 B.C.

From the Ahmes papyrus, we learn how easy addition is in the Egyptian number system. For example, to add 57 and 24

$$\cap \cap \cap \cap \cap \genfrac{}{}{0pt}{}{||||}{||||} \} \; 57$$

$$\cap \cap \; |||| \} \; 24$$

count 10 ones, convert them to a 10 (heel bone), and write the answer

$$\genfrac{}{}{0pt}{}{\cap \cap \cap \cap \; |}{\cap \cap \cap \cap} \} \; 81$$

One needs only to be able to count to 10 in order to add using the Egyptian number system. For example, ten ones are equal to one heel bone ($\cap$), ten heel bones are equal to one snare ($9$), and ten snares are equal to one lotus flower ($\maltese$).

The Egyptians developed a method of multiplication which is also reasonably easy, because it involves successive doubling which is, in essence, adding a number to itself. For example, to double 57 write

$$\cap \cap \cap \cap \cap \; \genfrac{}{}{0pt}{}{||||}{|||}$$

$$\cap \cap \cap \cap \cap \; \genfrac{}{}{0pt}{}{||||}{|||}$$

giving

$$9 \cap |||||$$

or 114, which is obtained by replacing 10 heel bones, $\cap$, by a snare, $9$, and ten ones by a heel bone, $\cap$.

Consider the example of 13 × 19. First write 19

$$(1 \times 19) \quad \genfrac{}{}{0pt}{}{|||||}{\cap \; ||||}$$

then double it

|||||
∩||||
|||||
∩||||

giving 38.

(2 × 19)  ∩∩∩ ||||
                    ||||

Double 38

∩∩∩ ||||
        ||||

∩∩∩ ||||
        ||||

giving 76.

(4 × 19)  ∩∩∩∩ |||
          ∩∩∩  |||

Double 76

∩∩∩∩ |||
∩∩∩  |||

∩∩∩∩ |||
∩∩∩  |||

giving 152.

(8 × 19)   ∩∩∩∩ ||

There is no need to double further, because 16 × 19 is greater than 13 × 19. The multiples of 19 needed to make 13 are 8, 4, and 1. Thus, add

**4  Early Mathematics**

8 × 19, 4 × 19, and 1 × 19 to find 13 × 19.

(8 × 19)   𝟫∩∩∩∩| |

(4 × 19)   ∩∩∩∩∩∩∩ | | |
                    | | |

(1 × 19)   ∩ | | | | |
             | | | |

giving 247.

(13 × 19)   𝟫𝟫∩∩∩∩ | | | |
                     | | | |

This procedure can be understood more readily by writing in our symbols as

| multiple of 19 | product |
|---|---|
| ✓ 1 | 19 |
| 2 | 38 |
| ✓ 4 | 76 |
| ✓ 8 | 152 |

Adding the products indicated, we find 152 + 76 + 19 = 247.

Division in the Egyptian number system is much more difficult, and before trying division it is essential to understand the Egyptian notion of a fraction. The Egyptians had only the concept of a unit fraction such as 1/3, 1/5, 1/27, 1/101. The fraction 1/4, for example, was called the fourth-part and written ⊏⊐ or in our symbols 4̄. Similarly, 1/10 would be ⋂̄ or 1̄0̄.

Think about fractions, and you can understand why it is plausible that only unit fractions should be conceived. As an example, cut a pie into fourths (fig. 1.1).

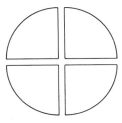

Figure 1.1

There are four separate pieces, each a fourth-part. No one piece is equal to 3/4 of the pie; all we observe are three fourth-parts. The Egyptians had no separate name or symbol such as 3/4 for this group of three objects.

The Egyptian concept of fractions, however natural it seems, made division very difficult, because it must be done by halving (multiplying by 1/2). To find 19 ÷ 13, for example, we calculate with 13 to obtain 19. (Recall that unit fractions will be represented by a bar over the numeral, i.e., $1/2 = \bar{2}$, $1/4 = \bar{4}$, etc.)

First, write 13.

$$(13 \times 1) \quad 13$$

Then halve 13.

$$(13 \times \bar{2}) \quad 6 + \bar{2}$$

Thirteen and 1/2 of 13 is greater than 19, so we must halve again, because we are looking for the multiples of 13 that will give us 19.

$$(13 \times \bar{4}) \quad 3 + \bar{4}$$

The sum of 13 and $13 \times \bar{4}$ is less than 19, so we need to add another value. We halve again.

$$(13 \times \bar{8}) \quad 1 + \bar{2} + \bar{8}$$

Adding $13 \times 1$, $13 \times \bar{4}$, and $13 \times \bar{8}$, we have $13 + 3 + \bar{4} + 1 + \bar{2} + \bar{8}$, or $17 + \bar{2} + \bar{4} + \bar{8}$, obviously not the 19 we seek.

It is easy to calculate with 13 to get 1; just take 1/13 of 13.

$$(13 \times \overline{13}) \quad 1$$

Adding to our previous sum, this gives $18 + \bar{2} + \bar{4} + \bar{8}$. We need another eighth-part to make 19. (If we did not know this from observation, we could use an Egyptian method of finding a common denominator, but this gets too complicated.) Since $\overline{13}$ of 13 gives 1, we see that successive halvings give

$$(13 \times \overline{26}) \quad \bar{2}$$
$$(13 \times \overline{52}) \quad \bar{4}$$
$$(13 \times \overline{104}) \quad \bar{8}$$

6   Early Mathematics

We see if we add 13 × 1, 13 × $\bar{4}$, 13 × $\bar{8}$, 13 × $\overline{13}$, and 13 × $\overline{104}$, we get 19. That is,

$$13 + 3 + \bar{4} + 1 + \bar{2} + \bar{8} + 1 + \bar{8} = 18 + \bar{2} + \bar{4} + \bar{8} + \bar{8} = 19$$

Therefore, adding the multiples of 13 from the above products should give us the answer we seek.

$$19 \div 13 = 1 + \bar{4} + \bar{8} + \overline{13} + \overline{104}$$

This is far from being a simple procedure. However, the Egyptians, over the course of the hundreds of years of making these calculations, learned the most efficient approach for each problem, and were far better at their division than the author.

Contrary to our previous statement, the Egyptians did have a symbol for a fraction which was not a unit fraction, 2/3. It was ⵡ, which meant 1/1(1/2). We write it as $\bar{\bar{3}}$. This symbol is used in another example of division, 2 ÷ 5.

(5 × 1)    5
(5 × $\bar{\bar{3}}$)    1 + $\bar{\bar{3}}$

Now only $\bar{3}$ is needed to make 2. If 5 × 1 gives 5, 5 × $\bar{5}$ gives 1, and 5 × $\overline{15}$ gives $\bar{3}$.

(5 × $\bar{5}$)    1
(5 × $\overline{15}$)    $\bar{3}$

Adding (5 × $\bar{3}$) + (5 × $\overline{15}$) gives 2, so the result of 2 ÷ 5 is $\bar{3}$ + $\overline{15}$.

Returning to the Ahmes papyrus, we find that four problems presented are to divide six, seven, eight, or nine loaves among 10 men. The answers are $\bar{2}$ + $\overline{10}$ each, $\bar{\bar{3}}$ + $\overline{30}$ each, $\bar{\bar{3}}$ + $\overline{10}$ + $\overline{30}$ each, and $\bar{\bar{3}}$ + $\bar{5}$ + $\overline{30}$ each, respectively. Checks are also given. The first calculation, 6 ÷ 10 (remember we are calculating with 10 to obtain 6), would be

        1    10
√    $\bar{2}$    5
√   $\overline{10}$   1   Result: $\bar{2}$ + $\overline{10}$

So each man gets a half loaf and a tenth loaf. (Notice that only the multiples of 10 are listed in the middle column. It is unnecessary to repeat 10 × each time.)

The Egyptian computation was applied to many problems such as determining the amounts of grain needed for making beer or bread and the finding of areas and volumes. For example, problem 51 of the Ahmes papyrus is to calculate the area of an isosceles triangle (fig. 1.2a).

You already know that the method of finding the area of a triangle is to take one-half of the base times the altitude. The interesting aspect is the justification of the method by showing how the isosceles triangle can be divided by the altitude into two right triangles which can then be joined to form a rectangle of height equal to the altitude and base equal to one-half the base of the triangle (fig. 1.2b). It is a kind of visual justification which the Egyptians employed.

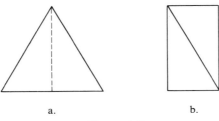

a.  b.

**Figure 1.2**

Problem 48 is to find the area of an octagon obtained from a square of side 9 by dividing each side into thirds and connecting the points (fig. 1.3). The method is to find the area of the square, $(9)^2 = 81$, and to subtract from that area the area of the four triangles in the corners. The area of each triangle is $1/2(3)(3)$, so the four triangles have area $4(9/2) = 18$. Thus, the area of the octagon is $81 - 18 = 63$. This octagon can be viewed as an approximation to a circle of diameter 9. Its area was found to be 63 which is almost that of a square of side 8. In problem 50 of the Ahmes papyrus the area of a circle of diameter 9 is calculated as that of a square of side 8, a method which may have been based on problem 48.

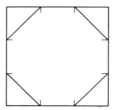

**Figure 1.3**

Problems 48 and 50 are geometric in nature. There are also problems in the Ahmes papyrus which we would class as algebraic. Problem 24 asks the value of a heap if a heap and one-seventh of a heap is 19. We would write this as $x + 1/7x = 19$, but the Egyptians used a method of solution later called the *rule of false position*. It involves making an intelligent guess for the heap, in this case 7, because it is easy to compute 1/7 of 7. Now, if a heap is 7, a heap and 1/7 of a heap is 8. But we want 19, so we multiply 7 by 19/8, arbitrarily putting the value we seek over the guessed value of the heap plus 1/7 of the heap. This procedure works since the problem is linear. The Egyptians divided 19 by 8 to get $2 + \bar{4} + \bar{8}$ and then multiplied by 7 to get an answer of $16 + \bar{2} + \bar{8}$.

There are other papyri of mathematical interest, but I will just mention one fragment which has a personal touch. The scribe who wrote it is teasing another scribe who came to him for help.

> You come to me to inquire concerning the rations for the soldiers, and you say "reckon it out!" You are deserting your office, and the task of teaching you to perform it falls on my shoulders.
> ... For see, you are the clever scribe who is at the head of the troops. A building is to be constructed with these dimensions [given]. The quantity of bricks needed for it is asked of the generals, and the scribes are all asked together without one of them knowing anything. They all put their trust in you and say, "You are the clever scribe, my friend! Decide for us quickly!" Behold your name is famous. Do not let it be said of you that there are things which even you do not know. Answer us how many bricks are needed for it?[1]

The scribe who wrote the papyrus then proceeds to work out the problem.

The Egyptians had a well-developed mathematical tradition and were capable of solving many useful mathematical problems. Their ideas and techniques influenced later generations of mathematicians. For example, unit fractions were used extensively, even as late as the Middle Ages. The doubling of a number, a basic step in the Egyptian method of multiplication, was one of the fundamental operations in many medieval arithmetic texts, although these texts also included more significant techniques learned from other peoples.

# 2 Babylonia

Mesopotamia is the region between the Tigris and the Euphrates rivers which is now known as Iraq. Since Babylon was such an important city

---

[1]Otto Neugebauer, *The Exact Sciences in Antiquity*, p. 79.

in that area, primarily from 2000 B.C. to 600 B.C., the name Babylonia is also often applied to the entire region.

Babylonian writing, called cuneiform (meaning wedge-shaped), is one of the oldest forms of writing, being dated from 4000 B.C. A rod, or stylus, was pressed into clay producing the wedge shapes, and the clay was then dried. This material was much more durable than the papyrus which the Egyptians used. In fact, in the ruins of ancient Babylonian cities these clay tablets were naturally preserved by being buried. As a consequence, many thousands of tablets have been recovered, only to decompose when exposed to the air. Otto Neugebauer, the scholar who first deciphered Babylonian mathematics in the late 1920s, points out the problem of preventing the discovered writings from decaying before they can be deciphered. The adventure involved in making expeditions to foreign countries to search for tablets often provides more excitement than trying to decode the cuneiform which has already been unearthed. It is no small wonder, therefore, that unstudied texts waiting to be deciphered in libraries and private collections are possibly suffering the destruction of decay.[2] Most of the recovered tablets come from the Old Babylonian period around 1800 B.C., the time of the great leader Hammurabi (ca. 2100 B.C.), or from the later Seleucid period (300 B.C.) named after one of Alexander the Great's generals.

In writing numbers, the Babylonians used, more or less, a sexagesimal system (base 60) but with only two symbols. The wedge in one position gives $\vee$ = 1; turned sideways it gives $\langle$ = 10. Any number up to 60 is written in a straightforward manner. For example, 32 = $\langle\langle\langle \vee\vee$. A place value system is used for larger numbers. Recall that our numeration system is also based on place value. When we write 143 we mean $1 \cdot 100 + 4 \cdot 10 + 3$. The value of each digit is determined by the place it occupies. We base our system on the number 10, while the Babylonians based theirs on 60. They wrote 143 as $2 \cdot 60 + 23$, or 2,23; in their symbols $\vee\vee \langle\langle \vee\vee$. Just as 59 is the largest number that the Babylonians could write using only one place, the number $59 \cdot 60 + 59$, or 59,59 (3599 in our system), was the largest they could write using only two places. To write 3600, they would write a one in the third place. This could be somewhat confusing, because the Babylonians did not use a zero sign until very late in their history. Whether $\vee\vee$ was 2 or $2 \cdot 60$, or perhaps $2 \cdot 3600$, had to be determined from the context. As a further example of a number expressed in their system, the number 61,802 which equals $17 \cdot 60^2 + 10 \cdot 60 + 2$, would be written as $\langle \vee\vee\vee\vee \langle \vee\vee$

This sexagesimal system was also used for fractions, and having studied Egyptian unit fractions we can appreciate how clever the Babylonians

---

[2]Ibid., p. 61.

were. They wrote fractions in descending powers of 60, just as we write decimal fractions in descending powers of 10. Thus, while 3.24 means $3 + 2(1/10) + 4(1/100)$ to us, the number ▽▽ ▽▽ ⟨▽ ▽ may have meant $6 + 12(1/60) + 3(1/3600)$ to the Babylonians. We can only guess, because the Babylonians never had the equivalent of a decimal point. In working with their system, we use a semicolon as a sexagesimal point, and commas between the other places for clarity, writing the above fraction as 6;12,3. Another example is $1/10 = 0;6$.

The advantage of this notation is that computations with fractions are done in the same way as computations with whole numbers, which is the same advantage that decimals have. In fact, the system was so much better for computations than other available systems, that it was used by astronomers long after the Babylonian period to calculate the positions of the moon and planets. This usefulness accounts for even our present use of this system. Our hour of time is still divided into 60 minutes and the minute into 60 seconds. The degree is also divided into minutes and seconds based on 60 units.

It is not known how the Babylonians came to develop their sexagesimal system, but several hypotheses have been put forth. Their system of measures was based on 60, so this may be one reason the method evolved.

$$1 \text{ talent} = 60 \text{ mana}$$

$$1 \text{ mana} = 60 \text{ bushels}$$

Just as we say "three forty-three" for three dollars and forty-three cents, they may have abbreviated 3 talents and 22 mana as 3,22. Also, there are tablets with two different stylus marks, a large $\vee$ = 60, $\langle$ = 10·60 and a small $\vee$ = 1, $\langle$ = 10. Perhaps the Babylonians eventually found it simpler to use one stylus and a positional system.

To familiarize ourselves with this base 60 system we shall do a few computations in an unhistorical manner, using our symbols and format. Note that just as tens are carried in our base system, sixties are carried in the Babylonian base system.

```
     21,49;17,42              3;12
   +  3,37;15,50              4;7
     ─────────               ─────
     24,86;32,92              21 84
   = 25,26;33,32            12 48
                            ──────
                            12;69,84
                          = 12;70,24
                          = 13;10,24
```

Babylonia

Of course, doing the multiplication involved using our memorized multiplication tables. The Babylonians also used multiplication tables, one of them being a nines table.

$$
\begin{array}{ll}
(9 \times 1) & 9 \\
(9 \times 2) & 18 \\
\quad \cdot & \\
\quad \cdot & \\
\quad \cdot & \\
(9 \times 20) & 3,0 \\
\quad \cdot & \\
\quad \cdot & \\
\quad \cdot & \\
(9 \times 50) & 7,30
\end{array}
$$

Some unusual multipliers are occasionally found in the clay tablets; there is a 7,30 table and a 44,26,40 table. What need is there for a 450 times table? This took scholars quite a while to solve, but they determined that 7,30 has to be interpreted as 0;7,30 which is 1/8. Thus, the 7,30 table is really a table for dividing by 8. The 44,26,40 table is for dividing by $81(1/81 = 0;0,44,26,40)$.

The Babylonians were much more advanced in mathematics than the Egyptians, and many of their tablets give solutions to algebra problems which we would formulate as equations. These cuneiform writings are, in fact, the earliest traces of numerical algebra. No explanations, only solutions, are given in the tablets. The solutions do exemplify general methods, however, as we will illustrate with an example. The area plus two-thirds the side of a square is 0;35. Find the side of the square. This can be formulated in modern terms as $x^2 + 0;40x = 0;35$ and a quadratic formula expression given for the solution.

$$x = \sqrt{\left(\frac{0;40}{2}\right)^2 + 0;35} - \frac{0;40}{2} = 0;30$$

This is an example of the quadratic formula

$$x = \sqrt{\left(\frac{p}{2}\right)^2 + q} - \frac{p}{2}$$

for the equation $x^2 + px = q$. This formula does not appear in the tablet, but you can see it being computed step-by-step. It is calculated in the following manner, using the value 0;40 which is two-thirds of one.

1. Half of 0;40 is 0;20.
2. Multiply 0;20 by itself (square 0;20), and the product is 0;6,40.

3. Add that product to 0;35, and the result is 0;41,40.
4. Take the square root of 0;41,40 which is 0;50.
5. 0;20, which was multiplied by itself, is subtracted from 0;50, and the result 0;30 is the side of the square.

If you follow each step you see that first the quantity under the square root sign is found, then the square root is taken, then the other term is subtracted from the square root, giving the final answer.

How might the Babylonians have obtained this method? We can only guess that the most plausible explanation is that they found it by actually completing the square. This is how they might have proceeded.

*Method*  First the given problem was probably represented geometrically. The area of the square, $x^2$, is represented by a square of side $x$, and $0;40x$ is represented by two rectangles, each of area $0;20x$ (fig. 1.4a).

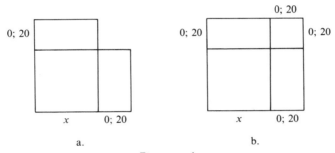

Figure 1.4

The condition of the problem is that the area of the square and the area of the two rectangles is equal to 0;35. Complete the square (fig. 1.4b) by adding an area of $(0;20)^2$ to the area of 0;35 to give a large square of area 0;41,40. The side of this large square is $\sqrt{0;41,40} = 0;50$. To get $x$ subtract the excess 0;20, with the result that $x = 0;30$.

You can see that the steps involved in completing the square are exactly those given in the procedure we listed.

The question of how the Babylonians computed the square root still remains to be answered, however. They used a method called the *divide and average* method which is now the one commonly taught in junior high schools. Their calculations involved sexagesimals, of course, but we will work in decimals. For example, to find $\sqrt{2}$ by the divide and average method, follow these steps:

1. Guess.
2. Divide the guess into the number under the radical, 2.
3. Average the two numbers resulting from the first two steps.
4. Use the average as a new guess, return to step 1 and repeat.

Suppose 1 was guessed in the first step.

| Guess | Divide | Average |
|---|---|---|
| 1 | $\dfrac{2}{1}$ | $\dfrac{1+2}{2} = \dfrac{3}{2}$ |
| $\dfrac{3}{2} = 1.5$ | $\dfrac{2}{1.5} = 1.33$ | $\dfrac{1.50 + 1.33}{2} = \dfrac{2.83}{2} = 1.415$ |

1.415 . . .

You see that our approximations are becoming closer to $\sqrt{2}$. This method of computing the square root has the advantage of being easy to remember, but can also be rather time-consuming. It works, because if the guess is smaller than the square root, dividing the original number by the guessed number will give a larger quotient than the square root, and the average of the guess and the quotient will be closer to the square root. Each repetition of the process produces an answer nearer to the square root. The Babylonians approximated to several places when the answer was not even.

Other problems presented in cuneiform tablets show that the Babylonians knew the formulas

$$(a + b)^2 = a^2 + 2ab + b^2$$

and

$$(a + b)(a - b) = a^2 - b^2$$

perhaps by means of geometric diagrams such as figure 1.5 which shows $(a + b)^2$ composed of the four areas, $a^2$, $b^2$, $ab$, and $ab$.

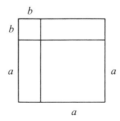

Figure 1.5

14 *Early Mathematics*

There are Babylonian tablets dealing with other geometrical problems, but I mention only one example — a tablet with a list of right triangles with integer sides. We know the easiest right triangle, the 3–4–5 (fig. 1.6),

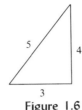

Figure 1.6

but few people will recall very many more. However, it is remarkably easy, using results available to the Babylonians, to obtain numbers $a$, $b$, and $c$, such that $a^2 + b^2 = c^2$. As the Babylonians were aware, such numbers can represent the sides of a right triangle. To find such numbers, the relationship

$$4xy + (x - y)^2 = (x + y)^2$$

with which the Babylonians were certainly familiar, can be used. If $4xy$ were a square, then the above relationship would show the sum of two squares equal to a third square, as required. To make $4xy$ a square, let $x = p^2$ and $y = q^2$ where $p$ and $q$ can be any integers. Substituting,

$$(2pq)^2 + (p^2 - q^2)^2 = (p^2 + q^2)^2$$

Thus, if $a = 2pq$, $b = p^2 - q^2$, and $c = p^2 + q^2$, the numbers $a$, $b$, and $c$ will represent the sides of a right triangle for any choice of $p$ and $q$. For example, if $p = 3$, $q = 2$, then $a = 12$, $b = 5$, $c = 13$, and one can check that indeed $12^2 + 5^2 = 13^2$. The Babylonian method was quite similar, but less general.

Returning to Babylonian algebra, we see that problems which we would now calculate by a unified method or formula were treated by the Babylonians (and by later peoples) by separate methods. For example, with our use of symbolism and our concept of negative numbers, we give one quadratic formula to solve all equations of the type $ax^2 + bx + c = 0$ for any $a$, $b$, or $c$, negative or positive. The natural approach of the Babylonians was to consider different appearing problems as separate cases, for example:

1. square = number       $x^2 = 9$
2. multiple of side = number     $5x = 8$

3. square + multiple of side = number     $x^2 + 3x = 5$
4. square = multiple of side + number     $x^2 = 3x + 5$
5. sum and product of two numbers given     $x + y = 5,$
                                                                                 $xy = 3$

After all, square = number is not the same problem as multiple of side = number, is it? The above five problems all look different and were treated as such by the Babylonians. To see the five problems as special cases of one general problem required great advances in concept and notation.

We have already seen the Babylonian step-by-step procedure for problems of the third type. Naturally, a different rule was used for each different type of problem; one of the fifth type would be handled in the following manner.

*Method*    The problem is to find $x$ and $y$ if $x + y = a$ and $xy = b$. One way of satisfying the first equation is to let $x = a/2$ and $y = a/2$, but this choice may not satisfy the second equation. Try to change $x$ and $y$ so they satisfy both equations. First, add a quantity $z$ to $a/2$ to give $x$ and subtract the same quantity from $a/2$ to give $y$. Then $x = a/2 + z$ and $y = a/2 - z$. The first equation is satisfied since

$$x + y = \frac{a}{2} + z + \frac{a}{2} - z = a$$

Now choose $z$ so that the second equation, $xy = b$, is satisfied.

$$xy = b$$

$$\left(\frac{a}{2} + z\right)\left(\frac{a}{2} - z\right) = b$$

$$\left(\frac{a}{2}\right)^2 - z^2 = b$$

$$z^2 = \left(\frac{a}{2}\right)^2 - b$$

$$z = \sqrt{\left(\frac{a}{2}\right)^2 - b}$$

Thus

$$x = \frac{a}{2} + z = \frac{a}{2} + \sqrt{\left(\frac{a}{2}\right)^2 - b}$$

and

$$y = \frac{a}{2} - z = \frac{a}{2} - \sqrt{\left(\frac{a}{2}\right)^2 - b}$$

This is the theory that underlies the Babylonian method of solution, and from this theory a step-by-step procedure identical to that used by the Babylonians can be developed.

1. Divide *a* by 2.
2. Square the result of step 1.
3. Subtract *b* from that square.
4. Take the square root of the quantity obtained by subtracting *b*.
5. Add the result obtained in step 4 to $a/2$ to get *x*.
6. Subtract this same value from $a/2$ to get *y*.

You might try this procedure on a numerical example to see how the Babylonian mathematicians followed these same steps in solving this type of problem.

The Babylonians had some well-developed mathematical procedures. Their level of mathematical achievement was the highest of the early civilizations, and many later peoples followed their lead in the development of rules for solving equations — algebra was born in Babylonia. Their positional system of numeration has been used to perform large numbers of computations, as in astronomy. In fact, although we did not discuss their astronomy in this text, the Babylonians became quite proficient at predicting the motion of the moon, including eclipses. The Greeks, whose mathematics we will study in chapter 2, learned much mathematics and astronomy from the Babylonians.

# 3 India

There was an early civilization in India, dating back to at least 3000 B.C. Written records from the early period are very scarce, however, because the Indians, unlike the Egyptians, did not build large tombs in which documents were preserved, and the writings were not protected by being buried, as was the case with the Babylonian tablets.

Early mathematics in India was closely related to religious rituals.[3] A work called *Sulvasutras* (Rules of the Cord) which contained many mathematical ideas was written in about 800–500 B.C., but may include

---

[3]See A. Seidenberg, "The Ritual Origin of Geometry," *Archive for the History of the Exact Sciences*, 1 (1961/62): 488–527.

material originally developed much earlier. There are indications of these ideas in the *Rig-Veda*, a collection of hymns dating from about 2000–1500 B.C.

In the *Sulvasutras* one finds rules for constructing right-angled altars using a cord of length 8 and two pegs placed 4 units apart (fig. 1.7). The cord is marked at a point 3 units from one end. The ends of the cord are tied to the pegs, *A* and *B*, the cord is stretched, and a peg is driven in at the marked point *C*. The result is a right angle at *B*. In the *Sulvasutras* constructions of straight lines and circles are done with pegs and cords. For example, a line can be constructed by stretching a cord between two pegs, or a circle can be constructed by using one stationary peg and a cord which is attached to the peg at one end, stretched, and rotated about the peg.

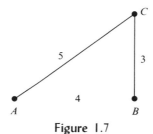

Figure 1.7

We will encounter ideas similar to these later in Greek mathematics, because various construction problems with which the Indians were concerned may have been the sources for Greek mathematics. For example, there was the problem of increasing the size of an altar while maintaining its shape. An altar in the shape of a bird was constructed to have $7\frac{1}{2}$ square units area (fig. 1.8).[4] The next step was to construct an altar in the same shape but with an area of $8\frac{1}{2}$ square units. A construc-

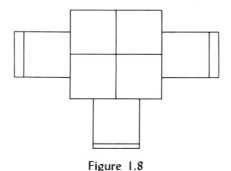

Figure 1.8

---

[4]Ibid., p. 491. See B. Datta, *The Science of the Sulba. A Study in Early Hindu Geometry* (Calcutta: University of Calcutta, 1932), 255 pp. for further details.

tion was developed which solves this problem, but we will not investigate it in this text.

Another problem given in the *Sulvasutras* was to construct a circular altar which had the same area as a given square altar. This is a very difficult problem whose converse later became famous in mathematics. An approximate solution given in the *Sulvasutras* follows:[5]

*Solution*  Construct the diagonals of the given square (fig. 1.9), and construct a circle with the diagonal as diameter and center $O$ where the diagonals intersect. Let $A$ be the midpoint of the side of the square. Construct $B$, the point at which the extension $OA$ intersects the circle, and construct $C$ such that $AC = 1/3 AB$. Then construct the circle with radius $OC$. This circle has area approximately equal to that of the given square.

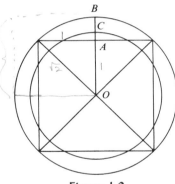

Figure 1.9

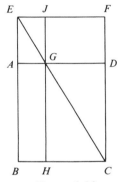

Figure 1.10

Constructing a rectangle equal in area to a given square is another problem mentioned in the *Sulvasutras*.[6] One side of the rectangle is given, but the solution is not expressed clearly in the *Sulvasutras*. A later commentator gives the following explanation in which the given side of the rectangle is assumed smaller than that of the square.

*Solution*  Given the square $ABCD$ (fig. 1.10), lay off the shorter side $HC$ of the rectangle. Complete the rectangle $HGDC$. Draw the diagonal $CG$ and extend it until it meets the extension of $AB$ at the point $E$. Complete the rectangle $BEFC$ and extend $HG$ to $J$ where it intersects $EF$. The rectangle $HJFC$ has the same area as the given square $ABCD$.

---

[5]Seidenberg, "The Ritual Origin of Geometry," p. 515.
[6]Ibid., pp. 517–18.

The proof is obtained by subtracting congruent triangles from congruent triangles:

*Proof*

$$\triangle EBC \cong \triangle EFC$$

but

$$\triangle EAG \cong \triangle EJG$$

and

$$\triangle GHC \cong \triangle GDC$$

Thus

$$\text{area } ABHG = \text{area } JGDF$$
$$\text{area } ABHG + \text{area } HGDC = \text{area } JGDF + \text{area } HGDC$$

Hence

$$\text{area } ABCD = \text{area } JHCF$$

We have seen something of the mathematics of three early civilizations, mathematics that developed from both practical and religious needs. Typically, the mathematics of these early civilizations appears as collections of problems and rules such as the Ahmes papyrus, the Babylonian clay tablets, and the Indian *Sulvasutras*. A significant feature of much of this early mathematics and an important legacy to future mathematicians was the use of numerical computation to solve varied types of problems. Even based on the incomplete records of the early mathematics that remain, we can conclude that it contained techniques and concepts of enduring usefulness.

## Problems

1. Add 287 and 464 using Egyptian symbols.
2. Multiply the following numbers in the Egyptian manner.
   a. $9 \times 15$        b. $23 \times 32$        c. $18 \times 27$
3. Divide the following numbers in the Egyptian manner.
   a. $9 \div 16$        b. $25 \div 4$        c. $14 \div 20$
4. If a heap plus an eighth of a heap gives 12, what is the value of a heap? Use the rule of false position.
5. Use the Egyptian method to find the area (approximate) of a circle of diameter 12.
6. Represent 2/7 as a sum of different unit fractions.
7. Represent 2/9 as a sum of different unit fractions in two different ways.
8. Using the Egyptian method, find how much each man would receive if
   a. seven loaves are divided among 10 men.
   b. eight loaves are divided among 10 men.
   c. nine loaves are divided among 10 men.
9. Problem 62 of the Ahmes papyrus asks for the amount of each precious metal in a sack which contains equal weights of gold, silver, and lead. The sack is bought for 84 sha'ty. A deben of gold is worth 12 sha'ty, a deben of silver worth six sha'ty, and a deben of lead worth three sha'ty. Solve the problem using the method of false position, first assuming that the sack contained one deben of each metal.
10. Represent the following numbers using Babylonian symbols.
    a. 172        b. 987
    c. 2371      d. 4000
11. Represent the following numbers using Babylonian symbols.
    a. 1/15       b. 5/12
    c. $7\frac{1}{10}$        d. 1/72
12. Find 1/7 correct to two sexagesimal places.
13. a. Which whole numbers $n$, from 2 to 10, give fractions $1/n$ whose decimal expansions are terminating? (An expansion is said to be

terminating if it has zeros from some point on. Thus $1/2 = .50$ is terminating while $1/3 = .33\ldots$ is repeating.)
   b. Which whole numbers $n$, from 2 to 10, give fractions $1/n$ whose sexagesimal expansions are terminating?
   c. Can you determine a rule which specifies which fractions will have terminating decimals? Sexagesimals?

14. Add 37,44;50,33 to 24,38;12,42.
15. Multiply 4;17 by 14;8.
16. Use the Babylonian method to find the following square roots. Continue until successive approximations agree to two decimal places.
    a. 7          b. 12          c. 20
17. The area of a square and 10 times its side are added, and the result is 24. What is the length of the side? Present your answer in a step-by-step form as the Babylonians did.
18. The area of a square and 0;30 times its side are added, and the result is 0;56,15. Find the side. Present your answer in a step-by-step form as the Babylonians did.
19. The sum of two numbers is 20 and their product is 36. Find the numbers. Present your answer in a step-by-step form as the Babylonians did.
20. Find four different right triangles (other than 3-4-5 and 5-12-13), each of which has as the lengths of its sides whole numbers with no factors in common with one another.
21. Show that the formula $(x + y)^2 = (x - y)^2 + 4xy$ can be obtained from the formula $a^2 - b^2 = (a + b)(a - b)$ by a suitable choice of $a$ and $b$.
22. Explain how figure 1.11 can be used to give a geometric representation of the formula $a^2 - b^2 = (a + b)(a - b)$.

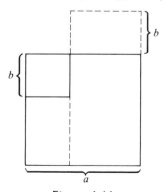

Figure 1.11

**23.** Given a square of side $s$, compute the area of the circle which can be constructed by the Indian method to be approximately equal to the area of the given square.


**References**

Introduction

    National Council of Teachers of Mathematics
    Seidenberg (1,2)                    Wilder

Egypt

    Boyer (1)                         Neugebauer (1)
    Gillings                          van der Waerden
    Midonick

Babylonia

    Aaboe                            Midonick
    Bruins                           Neugebauer (1)
    Gandz (1)                       van der Waerden

India

    Seidenberg (2,3)

# 2 Greek Mathematics

## Introduction

Greek civilization gave rise to excellent mathematics, philosophy, literature, and art. By 700 B.C. the Greek culture extended from mainland Greece to the islands in the Aegean Sea and to the west coast of what is now Turkey. Trade was expanding, and Miletus, a city on the coast of Turkey, was an important Greek trading center in the growing empire.

Literature began to change as the civilization became more progressive. Earlier, Homer and Hesoid described the origin of the world in human terms; the sky was a man and the earth a woman, etc. Nature itself was personified, and this incarnation pervaded the religion of the times. By 600 B.C., however, philosophers began to give natural explanations of creation. They suggested that everything was made of a combination of any or all of four basic elements — air, water, earth, and fire. It was thought that these basic substances, which symbolized states of matter, combined in different sequences and proportions to account for every detail in the universe. Our study will concern how mathematics was involved in this philosophical attempt to understand the world and how mathematics was connected with religion.

Unfortunately, there are almost no mathematical writings preserved from the period before 300 B.C., and even works written after 300 B.C. exist only as copies in much later documents. The Greeks wrote on papyrus, but they did not seal it in tombs as the Egyptians did. The only means of preservation was to copy writings repeatedly before the previous copy could disintegrate.

The earliest extant Greek work (except for a few fragments) is Euclid's *Elements* written about 300 B.C. Euclid's work became so popular that other writings ceased to be copied. He is so thorough in his compilation of elementary Greek mathematics, however, that scholars have been able to learn much about the period from 600 B.C. to 300 B.C. from the *Elements* without benefit of earlier texts. There are also scattered references to mathematics and mathematicians in other works by poets, historians, and philosophers. Although the total amount of material is not overwhelming, it does give some idea of the mathematics of the period. A commentary by a writer of A.D. 500 who had a copy of a history of mathematics written in about 320 B.C. is also a commonly used source of information, since the history itself has been long lost. The document is rather short, but does tell about the early Greek mathematicians.

The association of mathematics with philosophy led the Greeks to delve more deeply into the nature of mathematics than had earlier cultures. The Greeks wanted to know if a line could be measured exactly in terms of a given unit. They were not satisfied to obtain an approximate length, even though an estimation would have been sufficient for any practical use. The Greeks did approximate, particularly in the later period, but they always did so within the framework of a rich and well-described theory. Earlier cultures had general procedures but they never emphasized them. We are indebted to the Greeks for the development of explicit formal proof, and for the use of axioms.

# I  The Beginning of Greek Mathematics

## THALES (600 B.C.)

The first figure of importance in Greek mathematics is Thales of Miletus, but a few tales are all that is known about him. "Plato relates that he fell into a well while looking at the stars, and that a pretty Thracian slave-girl laughed at him, saying: 'he wanted to know what happens in the heavens, but he did not observe what was in front of his own feet!' "[1] His eyes were not always in the heavens, however. Miletus was a trading center, as we have mentioned, and Thales was a trader with good business

---

[1] B. L. van der Waerden, *Science Awakening*, p. 86.

sense. Aristotle tells of a good olive year in which Thales controlled all the oil presses and rented them at his own price.

Thales was among people who were thinking about the nature of the world. Being a trader, he had contact in his travels with Babylonian scholars. He probably organized some of the known facts of geometry which he learned from them, but he was not content with mere facts. Thales was among the first to be concerned with reasoning (so it is said). Among the few propositions that Thales is credited with are the following:

1. The angles at the base of an isosceles triangle are equal.
2. Any circle is bisected by its diameter.

What type of reasoning did Thales use? His notion of proof was probably that of intuitive "visualization," reminiscent of the Egyptian proof of the area of an isosceles triangle by breaking it into two right triangles and recombining them into a rectangle to "show" the method (see fig. 1.2). To give you another example of how Thales might have reasoned, I refer to Plato's Dialogue *Menon* in which Socrates taught an uneducated slave how to double the area of a square having sides 2 feet long, while maintaining the square shape.[2] (Socrates lived around 425 B.C., so this example is from a later period but still reflects an approach of Thales' time.)

First the slave thought to double the side, but Socrates drew such a square and showed the slave he was wrong (fig. 2.1a). The new area was four times that of the given square. The slave said the side must obviously be longer than the side of the original square in order to give a larger area, so he tried 3 as the new length, but this gave nine unit squares, not the 8 required (fig. 2.1b). Socrates then showed the correct method by doubling the sides and drawing a square to connect the midpoints of sides (fig. 2.1c). This new square contained four triangles while the original square contained only two, so it was the correct double. This

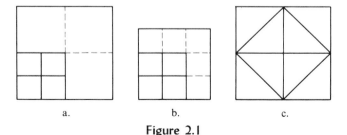

**Figure 2.1**

---

[2]Árpád Szabó, "The Transformation of Mathematics into Deductive Science and the Beginning of its Foundation on Definitions and Axioms," *Scripta Mathematica*, p. 35.

early type of proof by visualization is also apparent in Theorem I.4 of Euclid's *Elements* which we will study later in this chapter.

## PYTHAGORAS (ca. 540 B.C.)

Pythagoras of Samos was one of the most important figures in the history of science. He not only influenced Plato, but also other philosophers and scientists up to the time of the Renaissance. In his own time he was thought of primarily as a philosopher and religious teacher.

Pythagoras went to Egypt and Asia, learning about their cultures through his travels. He was aware of the many mystery rites, and he undoubtedly became familiar with religious rituals and their connection with number lore and geometrical rules. The cultures of these civilizations obviously impressed Pythagoras, and he assimilated some of their beliefs into his own philosophies while rejecting others.

He returned from his journeys with his newly acquired wisdom and founded a brotherhood of believers, the order of Pythagoreans, in Croton, a town on the east coast of Italy. In contrast to some Greek cults, which believed in ecstasy as a means of purification, the Pythagoreans believed that the contemplation of geometric form and numerical relations gave spiritual release. Music had always been thought to be a means of giving release, of purging the soul, and the Pythagoreans found that *numbers* underlay the harmony of music. For example, when a string is shortened to half its length, the tone produced when it is plucked is an octave higher. Similarly, the ratios 3 : 2 and 4 : 3 correspond to the harmonic intervals of the fifth and the fourth (fig. 2.2).

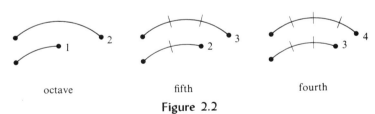

octave        fifth        fourth

**Figure 2.2**

The Pythagoreans felt that it was harmony which united the diverse aspect of segments into a whole. Harmony consists of numerical ratios. Thus, in a way, it is these numerical ratios of things which determine what they are and relate them to each other. That the Pythagoreans tried to understand the world through numbers is a very significant fact. Only in the Western world did modern science and mathematics develop, and the Pythagoreans initiated that development because of their beliefs concerning numbers.

Since Pythagoras was a famous man in his own time there were many stories told about him. His religious doctrines were ridiculed in some of the tales. When Pythagoras saw a dog being beaten he was supposed to have said, "Stop the beating, for in this dog lives the soul of my friend. I recognize him by his voice." Pythagoras' belief in the transmigration of souls was being spoofed. This was a serious belief of the Pythagoreans, however, and because of it they were vegetarians. Other anecdotes about Pythagoras afforded him quite a reputation. He was supposed to have been seen in two places at the same time, the calf of one of his legs was said to be made of gold, and most impressive, when he crossed a stream it reportedly rose up and greeted him saying, "Hail, Pythagoras."[3]

The Pythagoreans believed in the eternalness of mathematics, but abiding fellowship with the divine could only come after years of study and dedication.

> After a testing period and after rigorous selection, the initiates of this order were allowed to hear the voice of the Master [Pythagoras] behind a curtain; but only after some years, when their souls had been further purified by music and by living in purity in accordance with the regulations, were they allowed to see him. This purification and the initiation into the mysteries of harmony and of numbers would enable the soul to approach [become] the Divine and thus escape the circular chain of rebirths.[4]

The school of Pythagoras was active for over 100 years after his death. His followers did not generally use their own names on their work, thus we attribute the development of mathematics of this time to the Pythagoreans as a group, and we will refer to the concepts we shall discuss as Pythagorean. Their motto being "all is number," the Pythagoreans studied the various types of numbers such as even and odd numbers and perfect numbers. A perfect number is one which is equal to the sum of all its divisors except itself. There are not very many of these, the first two being

$$6 = 1 + 2 + 3$$

and

$$28 = 1 + 2 + 4 + 7 + 14$$

The next perfect number is 496.

---

[3]van der Waerden, *Science Awakening*, pp. 92–93.
[4]Ibid., p. 93.

Some of the Pythagorean concepts have been preserved in Euclid's *Elements*. Theorems 21–34 and Theorem 36 of Book IX have been shown to be the oldest parts of the *Elements* and are Pythagorean in origin.

**Theorem IX.21**  A sum of even numbers is even.

**Theorem IX.27**  Odd less odd is even.

These theorems culminate in Theorem IX.36.

**Theorem IX.36**  If $2^n - 1$ is prime, then $2^{n-1}(2^n - 1)$ is perfect.

| $n$ | $2^n - 1$ | $2^{n-1}(2^n - 1)$ |
|---|---|---|
| 2 | 3 | 6 |
| 3 | 7 | 28 |
| 4 | 15 (not prime) | |
| 5 | 31 | 496 |

It is still not known if there are infinitely many perfect numbers or not. It is also not known if there are odd perfect numbers. No one has ever found any, but no one has been able to prove that they do not exist.

Another source for Pythagorean ideas, especially number mysticism, is a book on numbers by a Neopythagorean, Nicomachus (A.D. 100). Numbers had symbolic significance, according to the Pythagoreans. For example, four is the number of justice or retribution, indicating the squaring of accounts. The Pythagoreans believed that all things could be explained by numbers; numbers were eternal while everything else was perishable. The philosophers before Pythagoras — Thales among them — emphasized the stuff from which the universe was made. The Pythagoreans emphasized form, proportion, and pattern, and their theories were based on their ponderings of what they observed. In fact, the word *theory* itself comes from *theoreo*, meaning *I contemplate*, and *thea*, meaning *spectacle*.

The Pythagoreans tried to express geometrical shapes as numbers. Believing that matter came in certain basic geometrical shapes and then combined to form things as we know them, they developed sequences in which a basic geometrical shape was represented by numbers, for example, the triangular numbers (fig. 2.3a) and the square numbers (fig. 2.3b). In fact we still call square numbers *square* because of the shape by which the Pythagoreans represented them. The Pythagoreans' attempt to explain geometry using numbers was not entirely successful, however, as we shall learn in the next section.

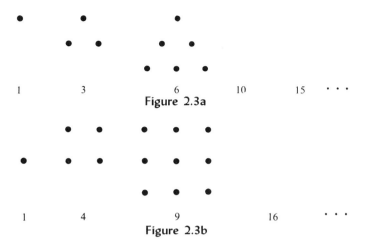

Figure 2.3a

Figure 2.3b

## 2  Crises and the Origin of Deductive Mathematics

The Pythagoreans applied their theories of numbers to points, lines, and planes. They made an assumption which appears very plausible, namely, that given any two lines there is a unit (perhaps small) which goes evenly into both of them an integral number of times. Thus the ratio of the lengths of lines could be expressed by numbers (see fig. 2.4 in which the ratio is 11 : 7). It was thought that for any two lines such a unit could be found, and two lines with this property are called *commensurable*.

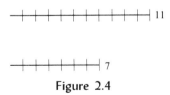

Figure 2.4

The Pythagoreans used the assumption of commensurability in their reasoning about geometry. In this way their philosophy that "all is number" was maintained, so it proved to be a great crisis for their school when they found that this theory was not true in all cases. Given a square (see fig. 2.5), the side and diagonal are incommensurable. It is not known exactly when this discovery took place, but it is sometimes mentioned that Hippasus (ca. 500 B.C.) was the one to disclose the fact of incommensurability and was supposedly set adrift at sea for revealing it.

The Pythagoreans were trying to keep the secret that the theory of commensurable lines was not true, but it is highly doubtful that they could not prove it false themselves. The incommensurability of the side of a square and its diagonal can be proved using the theory of even and odd numbers, a favorite study of the Pythagoreans, so the proof seems well within their grasp. The following example is an illustration of such a proof.

*Proof*  Assume there is a unit which goes into the side of a square $q$ times and the diagonal of the square $p$ times.

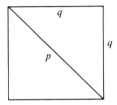

**Figure 2.5**

Then, by the Pythagorean theorem for isosceles right triangles,

$$q^2 + q^2 = p^2$$
$$2q^2 = p^2$$
$$\left(\frac{p}{q}\right)^2 = 2$$

Reduce $p/q$ to lowest terms so $p$ and $q$ have no factors in common. Since $2q^2$ is even by definition, and $2q^2 = p^2$, we know $p^2$ is even. The Pythagoreans were familiar with the theorem that if $p^2$ is even, then $p$ is even and can be represented as two times some other number, say $p_1$. Then $p = 2p_1$. Substitute, giving

$$2q^2 = (2p_1)^2 = 4p_1^2$$
$$q^2 = 2p_1^2$$

Now $2p_1^2$ is even, so $q^2$ is even, and therefore $q$ is even. But this implies $p$ and $q$ are both even and that is impossible because we reduced them to lowest terms. (If $p$ and $q$ are both even, they have a factor of 2 in common.) Thus, the hypothesis of commensurability leads to a contradiction.

The Pythagoreans had to reject many of their applications of numbers to geometry due to the proof of incommensurability. In particular, they had made much use of their theory of proportions, or ratios of numbers, which was probably much like what is found in Book VII of Euclid's *Elements*, with theorems such as $b : c = ab : ac$ and if $a : b = c : d$, then $ad = bc$. For example, they might have used numerical proportions in their theory of similar triangles. Take similar triangles and assume there is a unit going evenly into all sides (fig. 2.6). Then the corresponding sides are proportional in a numerical sense $q'/q = p'/p$.

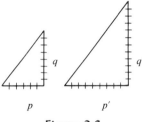

Figure 2.6

This theory of proportions of similar triangles may have been the original Pythagorean approach to the proof of the Pythagorean theorem, which states that in any right triangle the square on the hypotenuse is equal to the sum of the squares on the sides. The following example is a guess as to how that went.[5]

*Proof*  Let a right triangle $ABC$ be given and let squares be constructed on its sides and hypotenuse (fig. 2.7). Construct the perpendicular $OF$ to $BA$. Find the greatest common measure of the four lines $BC$, $CA$, $BO$, and $OA$ (assuming commensurability). In terms of this length as a unit, let the four lines be of length $a$, $b$, $d$, and $c - d$, respectively. Since angle $COB$ and angle $ACB$ are both right angles

$$\angle COB = \angle ACB$$

Also,

$$\angle CBO = \angle CBA$$

---

[5]See Sir Thomas L. Heath, trans., 3 vols., *The Thirteen Books of Euclid's Elements* (New York: Dover Publications, 1956) 1: 352-55, for a discussion of early proofs of this theorem.

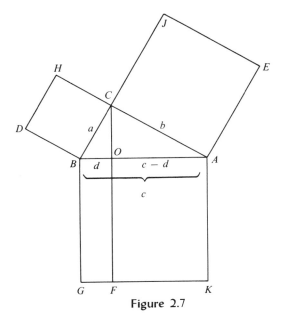

**Figure 2.7**

Therefore,

$$\triangle CBO \sim \triangle ABC$$

Using ratios of corresponding sides,

$$\frac{c}{a} = \frac{a}{d}$$
$$a^2 = cd$$

Thus, the area of the square $CBDH$ is equal to the area of rectangle $BOFG$. In a like manner, the area of square $CAEJ$ is equal to the area of rectangle $OAKF$. Since $BOFG + OAKF = BAKG$, then $a^2 + b^2 = c^2$ by substitution. The theorem follows that the square on the hypotenuse equals the sum of the squares on the sides.

Here you see use being made of the theory of numerical proportions based on the false assumption that any two lines are commensurable. Later we will see correct proofs of the Pythagorean theorem which are given in Euclid's *Elements*.

Although the Pythagoreans had to discard any proofs they made using the false assumption of commensurability, they undoubtedly developed many valid theorems of geometry. Unfortunately, very few references to their results have survived. Those that do indicate that their geometry, now considered as a separate subject not based on numbers, contained some of the theorems found in the first two books of Euclid's *Elements*.

The discovery of incommensurability caused quite a shock. Perhaps other results which appeared intuitive by visualization (such as commensurability) were faulty. The mathematician learned to be more careful and to depend more on reasoning and logic than on diagrams and displays. We shall see that at several points in history crises such as the proof of incommensurability caused mathematicians to be more careful and rigorous in formulating their theories.

## ZENO (ca. 450 B.C.)

Some scholars believe that Zeno had great influence on mathematics with his very ingenious arguments. Zeno was a disciple of Parmenides of Elea whose followers formed a school of philosophers, the Eleatics, who were critical of the Pythagoreans to some extent. Zeno gave brilliant arguments criticizing views of other philosophers, showing that the other person's hypothesis led to two contradictory conclusions. Thus, the hypothesis itself was impossible.

His paradoxes were quite amusing and have influenced thought to this day. Consider, for example, the race between Achilles and the tortoise. Achilles is faster so he starts at $A_1$ while the tortoise has a head start and begins at $T_1$ (fig. 2.8). When Achilles reaches the position where the tortoise started, the tortoise has moved on, not too far of course, but the tortoise is still ahead. When Achilles reaches this new position of the tortoise, $T_2$, the tortoise has moved still further to $T_3$. Thus, Achilles will never catch the tortoise.

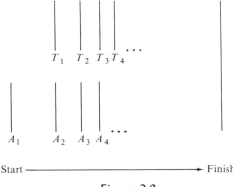

Figure 2.8

Zeno's argument seems theoretically convincing, although the tortoise would never win in reality. Modern mathematicians have discussed this paradox, considering the time it took Achilles to make an infinite number of these steps. The points Zeno raises about the divisibility of space and time are quite complex and philosophical. Is there a smallest unit of space? Is there an instant — a smallest unit of time?

I would like to explain one of Zeno's points in a mathematical context. Suppose we choose two numbers, 3 and 6 (fig. 2.9).

Figure 2.9

It is clear, and no one argues otherwise, that the whole, 6, is greater than the part, 3. Now consider two line segments (fig. 2.10) and suppose they

Figure 2.10

are composed of points. A one-to-one correspondence can be set up between points of the first line and points of the second (fig. 2.11).

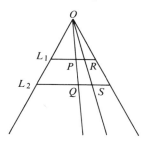

Figure 2.11

We connect the right endpoints and the left endpoints of lines $L_1$ and $L_2$ and extend these two lines of connection until they intersect at $O$. Now, given any point $P$ on $L_1$, draw the straight line from $O$ through $P$ and extend it until it crosses $L_2$ in $Q$. Therefore, we say $Q$ corresponds to $P$. Similarly, given any point $S$ on $L_2$ we find the point $R$ on $L_1$ which corresponds to it. Thus, there are just as many points on $L_1$, the part, as there are on $L_2$, the whole, because each point on $L_1$ is paired with a unique point on $L_2$ by this process. Not only could the Pythagoreans not prove that the whole line was greater than the part, but Zeno's argument, similar to the one above, seemed to show that the whole is equal to the part.

A new crisis had arisen for the Pythagoreans, because another one of their basic notions had apparently been proven false. They could not consider a line to be made up of points, because then they would have to respond to the argument that a shorter line having just as many points as a longer line was, in fact, equal to it. Rather than give up the study of geometry, the Pythagoreans refused to consider a line as made up of points, and they accepted the common idea that the whole is greater than the part. We find this statement later as a *common notion*, or *axiom*, in Book I of Euclid's *Elements*. The other common notions also concern equality. It seems that the Eleatics' arguments influenced the Pythagoreans to adopt axioms rather than abandon their ideas. Therefore, a possible explanation for the development of propositions which were accepted as self-evident truths is that the Eleatics forced the Pythagoreans to adopt as axioms statements in which they believed despite the arguments of the Eleatics against them.[6]

We have learned about Pythagorean ideas of numbers and philosophy and how the discovery of incommensurables made them discard some geometry based on numbers. By 450 B.C. criticism of the nature of points may have led to the use of axioms, certainly an important occurrence in the history of mathematics. Before continuing with an explanation of Greek geometry, however, it will be of benefit to study Greek numerals.

## 3  Greek Number Systems

There were two systems of Greek numerals, and both were developed at approximately the same time between 800–500 B.C. The Attic is the more primitive and is named for Attica, or Athens. The other system is Ionic, named for the region of Ionia on the coast of Asia Minor (Turkey).

### ATTIC GREEK

In this system the first letter of the word for five was used as its symbol, just as if we were to use $f$ as the symbol for the number five. Strokes were

---

[6]See Szabó, "The Transformation of Mathematics," pp. 27–48, 113–39, for the development of this argument.

used for 1, 2, 3, and 4 — |, | |, | | |, | | | |. Other numbers were symbolized in the following manner:

    Π or Γ (an old form of π)   the letter pi, initial of ΠΕΝΤΕ, five, is used as a numeral for 5.

    Δ   the letter delta, initial of ΔΕΚΑ, ten, used as a numeral for 10.

    H   like our *h*, initial of HEKATON — 100.

    X   chi, initial of XIΛIOI, chil'ioi — 1000.

    M   mu, initial of MYPIOI, myrioi — 10,000.

Position was not important in the Attic system. A relatively large number of symbols were required to write a number, for example

$$473 = \text{HHHH } \Gamma \Delta\Delta \mid\mid\mid$$

Here the ᛮ represented five 10's, or 50, the Γ being 5 and the Δ being 10.

## IONIC GREEK

The Ionic Greek was an alphabetical numeral system; the letters of the alphabet were used as symbols. There was an older form in which the 24 Greek letters represented the numbers from 1 to 24 but this was not very convenient for computations or for writing numbers larger than 24.

The Ionic system used 27 letters. The Greek alphabet, derived from the Phoenician, was augmented by three archaic letters. The alphabetical representations of the Ionic system are given in the following chart. (For ease of recognition the letters are written in lowercase, although the Greeks of that period used uppercase.)

| 1 | α | 10 | ι | 100 | ρ |
|---|---|----|---|-----|---|
| 2 | β | 20 | κ | 200 | σ |
| 3 | γ | 30 | λ | 300 | τ |
| 4 | δ | 40 | μ | 400 | υ |
| 5 | ε | 50 | ν | 500 | φ |
| 6* | ς | 60 | ξ | 600 | χ |
| 7 | ζ | 70 | ο | 700 | ψ |
| 8 | η | 80 | π | 800 | ω |
| 9 | θ | 90* | ϙ | 900* | ϡ |

*The letters for 6, 90, and 900 were the archaic letters.

38   Greek Mathematics

Position was not important in the Ionic system either, but the Ionic System was more economical than the Attic, because fewer symbols were required to represent a number. For example,

$$573 = \overline{\phi o \gamma}$$

Only two symbols are needed to write 570.

$$570 = \overline{\phi o}$$

Bars were written over the numerals to distinguish them from words.

The Greeks were able to extend the Ionic system to represent large numbers. The symbols for 1 to 9 were written with a left subscript to represent 1000 to 9000.

| $_{\prime}\alpha$ | $_{\prime}\beta$ | $_{\prime}\gamma$ | $_{\prime}\delta$ | . . . |
|---|---|---|---|---|
| 1000 | 2000 | 3000 | 4000 | |

For numbers larger than 10,000 the symbol M for myriads was used. Thus 30,000 could be written as 3 myriads or $\overset{\gamma}{M}$. Another example is

$$\overset{\lambda\beta}{M}_{\prime}\alpha\sigma\pi\gamma = 321{,}283$$

Addition and multiplication were difficult in the Ionic system, but the system was very convenient for writing numbers in order to keep records. A late nineteenth century historian, Tannery, said that he had practiced addition and multiplication in this system and found them quite reasonable. Even still, the average tradesman in the days of the Greeks probably used an abacus to add bills and keep accounts.

## 4  Early Results and Problems of an Independent Mathematics

Returning to Greek geometry, we recall that congruences, parallels, and triangles had been studied by the Pythagoreans and some basic theorems of geometry had been developed. These theorems will be described in the later form in which we find them in Euclid's *Elements*.

# THE THREE FAMOUS PROBLEMS

By about 450 B.C. scholars were beginning to study more advanced problems. Recall that in India, peg and cord constructions were used to draw lines and circles. Perhaps this influenced Greek mathematicians to postulate the construction of lines and circles.[7] In any case, three such basic rules of construction were specified and later found as Euclid's first three postulates. They are the following:

**Postulate I.1**    To draw a straight line between two points.

**Postulate I.2**    To extend any line indefinitely.

**Postulate I.3**    To draw a circle with any center and radius.

Many theorems are proved using these construction postulates, and all such constructions are possible using a straightedge and compass. The Greeks now were determined to actually construct (at least theoretically) the solution to a problem. They did not want to be proved wrong again, as with incommensurability.

As an example of a construction which is possible using a straightedge and compass, consider the bisection of a line segment $AB$ (fig. 2.12).

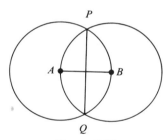

Figure 2.12

1. Draw a circle with center $A$ and radius $AB$ (postulate 3).
2. Draw a circle with center $B$ and radius $AB$ (postulate 3).
3. Connect the points where the circles intersect, $PQ$ (postulate 1).
4. The segment $PQ$ bisects $AB$, as is proved in *Elements* I.10.

Here Euclid assumed it was clear that the circles actually intersected in points $P$ and $Q$.

---

[7]See A. Seidenberg, "Peg and Cord in Ancient Greek Geometry," *Scripta Mathematica*, pp. 107–22.

There were, however, three problems which became well known because no one could solve them with a straightedge and compass. These are sometimes called the *three famous problems*. They are:

**Problem 1**  To trisect any angle (with straightedge and compass only).

**Problem 2**  To construct a square with area equal to that of a given circle (called *squaring the circle*).

**Problem 3**  To construct a cube with volume double that of a given cube (called *duplication of the cube*).

It is possible to solve these problems using *more* than just a straightedge and compass. Later we will see an example of a trisection. It was not until the 1800s that these problems were shown to be impossible to solve using only a straightedge and compass, so all the efforts over 2200 years were to no avail. It is interesting to note that the methods of algebra were used to show the impossibility of these geometric problems.

Perhaps these problems were of earlier origin than Euclid's time.[8] Both the Egyptians and the Indians gave approximate solutions to problem 2. Recall that in India altar construction led to mathematical problems, and there is a Greek legend about the origin of the duplication of the cube problem which relates it to altar construction. The story has it that it was announced to the people of Delos through an oracle that in order to be liberated from the plague they would have to make an altar twice as great as the existing one to Apollo. The architects were much embarrassed in trying to find out how one solid could be made twice as great as another one, so they asked the mathematicians. Thus, the problem of the doubling of the cube was posed, but not solved.

## HIPPOCRATES (ca. 430 B.C.)

One person who lived in this time and who worked on the three famous problems was Hippocrates of Chios (not the Hippocrates of the Hippocratic oath who was a doctor from Cos). In trying to square the circle, Hippocrates showed the area of the shaded *lunules* was equal to that of the shaded isosceles right triangle (fig. 2.13). He hoped to extend this result to find the area of a whole circle equal to a square, but, as we now know, that is impossible.

---

[8]See Seidenberg, "The Ritual Origin of Geometry," pp. 493–94.

*Early Results and Problems of an Independent Mathematics*

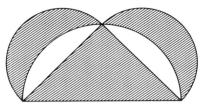

Figure 2.13

Hippocrates wrote the first known compilation of *Elements of Geometry*. The text did not survive, but it probably contained elementary geometry of triangles and parallels as was known to the Pythagoreans and found in Book I of Euclid's *Elements*, as well as Hippocrates' own work on circles and polygons which can be found in Books III and IV of Euclid's *Elements*. We will discuss some results on polygons which were known by the time of Hippocrates.

An equilateral triangle, a square, and a regular pentagon can be inscribed in a circle (fig. 2.14) with a straightedge and compass. If each

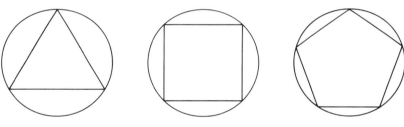

Figure 2.14

side of the equilateral triangle is bisected and the corresponding points on the circle connected, then a regular hexagon inscribed in the circle is obtained. Similarly, by bisection a 12-sided polygon can be constructed from the hexagon (fig. 2.15). Thus, given that any one polygon is constructible, a polygon with double the number of sides of the given polygon can be constructed. From these three polygons — the equilateral

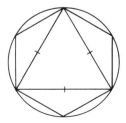

Figure 2.15

triangle, the square, and a regular pentagon — we know that polygons of the following numbers of sides are constructible.

$$3, 6, 12, 24, 48, \ldots$$
$$4, 8, 16, 32, 64, \ldots$$
$$5, 10, 20, 40, 80, \ldots$$

The Greeks of this time also could use a triangle and a pentagon together to construct a 15-sided polygon giving the constructible sequence

$$15, 30, 60, 120, \ldots$$

But there were regular polygons of 7, 9, 11, 13, 14, 17, and more sides that no one was able to construct with a straightedge and compass. In 1796 the genius Gauss, then 19, proved that the 17-sided polygon was constructible. (Here, again, it is interesting to note that the proof was accomplished using algebra.) In fact, Gauss developed a rule (which we will study later) for finding other constructible polygons. The next new constructible polygon of a prime number of sides is 257 sided. The 17-sided polygon is fairly difficult to construct, but someone has figured out and carried through the construction of the 257-sided polygon which is immensely difficult, and someone even spent years constructing the next prime-sided polygon which has over 65,000 sides!

We will investigate a particularly easy construction of a regular pentagon attributed to Ptolemy, who lived around A.D. 150 (fig. 2.16).

1. Construct a circle with perpendicular diameters.
2. Bisect $DC$, giving $E$.
3. Construct a circle with $EB$ as radius and $E$ as center which cuts $AD$ at $F$.
4. Connect $BF$. $BF$ is the side of a regular pentagon. This can be shown by laying it off around the circle. A proof is outlined in problem 15.

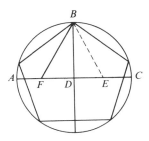

Figure 2.16

*Early Results and Problems of an Independent Mathematics*

Pentagons had been very interesting to the Pythagoreans, and no doubt some of the results concerning constructibility of polygons were known to them. In fact, the *pentagram* (five-pointed star) (fig. 2.17) was a mystic symbol of the Pythagoreans. There is a story about a Pythagorean in a foreign country who was taken care of until his death by a kind man. The Pythagorean was unable to pay the man, but from his death-bed, he instructed the man to paint a pentagram on the outside wall of his house, so that any Pythagorean who might pass by would come in and visit. A Pythagorean did come past many years later and rewarded the man for his kindness.

Figure 2.17

The pentagram is a rather interesting figure. Inside the star is a pentagon. If the vertices of this pentagon are connected, another star is formed. Inside that star is another pentagon which contains another star which contains another pentagon, ad infinitum (fig. 2.18a). The pentagram is

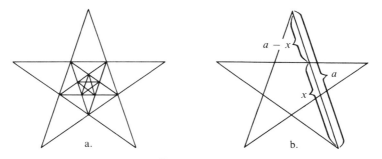

Figure 2.18

also related to what was later known as the *golden section*, so called because it was thought to be a pleasing division of a line into two parts. The golden section, or *golden mean*, has the property that one side of the star, $a$, is divided by the other side into two parts, $x$ and $a - x$ (fig. 2.18b), such that

$$\frac{a}{x} = \frac{x}{a - x}$$

## MATHEMATICS IN GREEK EDUCATION

Greek schools were all private and open only to male students. Women were educated in the home, chiefly in the domestic arts. Arithmetic was taught to Greek students (along with reading, music, gymnastics, writing) until they reached age 14. Their mathematical activities included finding areas of boxes, playing counting games, and distributing apples, among other exercises. Geometry and astronomy were taught to ages 14–18 during secondary education.

Isocrates, who wrote on mathematics in Greek education, said that if it does no other good, the teaching of mathematics keeps the young out of mischief and the study of it helps to train one's mind and sharpen one's wits.[9] So it seems, the justification of mathematics education was much the same as it is at present.

## 5 New Methods and Ideas

We come now to a period 40 or 50 years after the time of Hippocrates when several problems which had been attacked unsuccessfully earlier were finally solved. Eudoxus showed how to work with ratios of magnitudes correctly and also solved difficult area and volume problems, which we would now compute using calculus. These new Greek methods are more subtle than any we have previously encountered.

Mathematics at this time was viewed very broadly. The word *mathematics* itself comes from the Greek word *mathemata*, meaning *things learned* or *subjects of instruction*. Around 390 B.C. those subjects were geometry, arithmetic, music, and astronomy. Other practical skills were probably learned in a less formal manner.

**Archytas of Taras (ca. 390 B.C.)** was a Pythagorean and was probably the man who introduced Plato to mathematics. Recall that music was understood using ratios or proportions of numbers, the 2:1 ratio being an octave, etc. Archytas studied the theory of ratios and proportions in relation to his theory of music and wrote a book on music and harmony. Much of Book VIII of Euclid's *Elements* on proportions may be attributable to Archytas.[10] Archytas also developed an ingenious solution to the problem of the duplication of the cube, but it is extremely complicated.

---

[9]Sir Thomas L. Heath, *Greek Mathematics*, pp. 7–8.
[10]See van der Waerden, *Science Awakening*, p. 153.

## PLATO (380 B.C.)

Plato of Athens is one of the most famous Greek philosophers. He did not do any noteworthy work with mathematics, but he was affected by the Pythagorean number philosophy and, in turn, incorporated it into his work, influencing later generations. Most of Plato's works have survived, and he is the first Greek philosopher to have more than a minute portion of his writings preserved. Plato wrote on almost every subject, so later generations have looked to him for an extensive view of Greek thought. He was also very influential in his own time and founded the Academy on the outskirts of Athens which was to continue for 900 years until A.D. 529. Over the door was the motto, "Let he who is ignorant of geometry not enter here."

Plato believed that the essence of reality was the eternalness of geometric forms and numerical relations, in contrast to the transitoriness of material things. He wished to purify the soul by contemplating the eternal and developed a theory of ideal forms. The study of mathematics was recommended because it dealt with eternal geometric forms such as circles and triangles. The perfect circle of mathematics is an idea which is only imperfectly represented by inaccurate drawings in sand or on paper. The senses were imperfect, and some ideas such as incommensurability which could only be grasped by indirect reasoning were not apparent to the senses from a picture. The ideal forms were perfect and eternal, unlike the imperfect senses. This philosophy of mental contemplation played an important role in the development of mathematical science. By contrast, other cultures depended more on sensual evidence.

The Pythagorean ideas on regular polyhedra (solids formed by intersecting planes) were adopted by Plato. There are only five regular polyhedra — polyhedrons which have congruent faces and equal angles at every vertex. (With two-dimensional figures, however, there is a regular polygon of $n$ sides for every integer $n \geq 3$.) The five regular polyhedra are the following:

| Polyhedron | Number of faces | Shape of faces |
|---|---|---|
| tetrahedron | 4 | triangles |
| cube | 6 | squares |
| octahedron | 8 | triangles |
| dodecahedron | 12 | pentagons |
| icosohedron | 20 | triangles |

It is easy to understand why there are only five regular polyhedra. If you flatten any corner of any polyhedron the sum of the angles of the polygons joined at that point will be less than 360°. Consider the possibilities for joining regular polygons. Of course, we need at least three

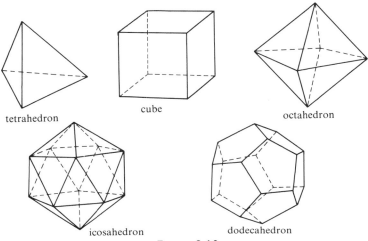

**Figure 2.19**

faces joined at each corner to make a solid. Each angle of an equilateral triangle contains 60°. We can join the following at each corner:

| Number of equilateral triangles | Sum of the angles | Polyhedron formed |
| --- | --- | --- |
| 3 | 180° | tetrahedron |
| 4 | 240° | octahedron |
| 5 | 300° | icosahedron |
| 6 | 360° | none possible |

A square has four angles of 90°, so we can join the following at each corner:

| Number of squares | Sum of the angles | Polyhedron formed |
| --- | --- | --- |
| 3 | 270° | cube |
| 4 | 360° | none possible |

A pentagon has 360° + 180° = 540° for its five angles, so each angle is 108°. Thus, joining three pentagons would produce a sum of 324° at each corner, and the polyhedron formed is a dodecahedron. A hexagon has 540° + 180° = 720° for its six angles, so each angle is 120°. Joining three hexagons, the sum of the angles would be 360°, so no polyhedron is possible with hexagonal faces. Similarly, no polyhedron is possible with faces of seven sides or more.

In the *Timaeus* Plato states a theory whereby the basic geometrical forms, triangles, combine to make up the basic elements in the forms of

**New Methods and Ideas** 47

the regular polyhedra. Recall that the Greeks believed there were only four basic elements — fire, earth, air, and water. Plato related the elements to four regular polyhedra in the following manner:

$$\text{fire} = \text{tetrahedron}$$
$$\text{earth} = \text{cube}$$
$$\text{air} = \text{octahedron}$$
$$\text{water} = \text{icosahedron}$$

That is, fire was thought to be made up of particles of tetrahedral shape, and so on. To include the fifth regular solid, Plato made it a symbol of the world.

$$\text{world} = \text{dodecahedron}$$

We now have our modern atomic theory which does have some similarities to Plato's theory in that we assign basic structures to molecules and crystals. Later we shall see that Plato's ideas did, in fact, inspire Kepler.

Plato wrote a dialogue about **Theaetetus of Athens (ca. 375 B.C.)** in which he said Theaetetus proved that a line whose square $N$ is not in one of the ratios $1:1, 4:1, 9:1, \ldots$ with the unit square is incommensurable with the unit. Thus the side of the first square in figure 2.20 is incommensurable with the side of the second. In modern terms, $\sqrt{N}$ is irrational if $N$ is not a perfect square.

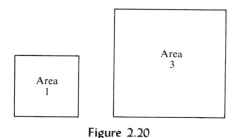

Figure 2.20

To prove this proposition Theaetetus greatly developed the theory of incommensurable line segments which is found in Book X, a very long volume of Euclid's *Elements*. This theory was applied to study the nature of the sides of regular polyhedra. In writing Book XIII of the *Elements* which is about regular polyhedra, Euclid may have drawn on the work of Theaetetus. We can see where the irrational would enter as, in modern terms, $1/4\sqrt{10 - 2\sqrt{5}}$ is the side of a regular pentagon inscribed in a circle of diameter one. There is some question as to the contributions of Theaetetus to Euclid's *Elements*. It is an interesting, if not impossible, problem to try to discover which persons are responsible for parts of Euclid's *Elements*.[11]

The period around 375 B.C. in which Theaetetus lived was marked by the constant advancement and sophistication of mathematics. Mathematicians continued to improve the work of their predecessors, as we will see with Eudoxus.

## EUDOXUS (ca. 370 B.C.)

Recall that with the discovery of incommensurability lines could no longer be assumed to have an integral ratio to one another. Given two lines $A$ and $B$ we may not be able to find two numbers, such as 10 and 13,

$A$ _____ $C$ _____

$B$ _____ $D$ _____

Figure 2.21

so that line $A$ is to line $B$ as 10 is to 13. Yet, clearly there is some length relationship between $A$ and $B$. We can see, for example, that $A$ is smaller than $B$. Given another pair of lines $C$ and $D$, we can see that the ratio of line $A$ to line $B$ is greater (in this case) than the ratio of line $C$ to line $D$ (fig. 2.21).

One achievement of Eudoxus of Cnidus was to show how to work with ratios of lines without using the false assumption of the commensurability of any two lines. Instead of dividing the lines into parts, he multiplied the lines. Thus, we find no theory of divisibility in geometry as we do in number theory. Eudoxus developed the theory of ratios of magnitudes. The result is found as Book V of Euclid's *Elements*. Eudoxus did not comment on ratio itself; he simply presented it for observation (fig. 2.21). He did, however, give a definition of when two ratios are equal (definition V.5) and a definition of one ratio being greater than another. A paraphrase of his definition follows.

**Definition V.7** Given two ratios, $A : B$ and $C : D$, the ratio $A : B$ is greater than the ratio $C : D$ if there are integers $m$ and $n$ such that $mA > nB$, but $mC \leq nD$.

---

[11]In addition to research papers, the book *Science Awakening* by B. L. van der Waerden contains much of this type of analysis.

We will apply this definition to a simple example (fig. 2.22).

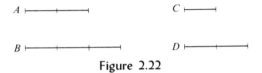

Figure 2.22

**Example** Choose, by trial and error, $m = 2, n = 1$. Then $mA > nB$, but $mC = nD$, so the definition of $A : B > C : D$ is satisfied (fig. 2.23).

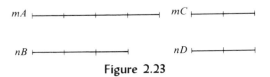

Figure 2.23

Using the same multiples, $m$ and $n$, we can get $mA$ greater than $nB$, but at the same time $mC \leq nD$. Thus, it seems reasonable that $A$ is bigger relative to $B$ than $C$ is relative to $D$.

As you can see, this is a difficult concept to understand, and it demonstrates the higher level of reasoning that the mathematics of Eudoxus required.

Eudoxus also made great advances in finding areas and volumes. Mathematicians of the early civilizations of Egypt and Mesopotamia could find the areas of rectangles, triangles, and other figures bounded by straight lines. They could not, however, find the area of a region such as a circle which is bounded by a curved line. Eudoxus developed a very clever method for solving some of these harder problems. His method involved some kind of a limiting process and was similar in that respect to calculus. We will examine a couple of Eudoxus' ideas, all of which are developed in Book XII of Euclid's *Elements*.

For example, Eudoxus proved that the areas of two circles (fig. 2.24) are in the ratio of the squares of their diameters,

$$\frac{A_1}{A_2} = \frac{d_1^2}{d_2^2}$$

It had been known for a long time that if similar polygons are inscribed in circles, their areas are in the same ratio as that of the squares of the diameters of the circles (fig. 2.25). If these polygons are constructed with more and more sides, they approach the circle. Since the theorem is

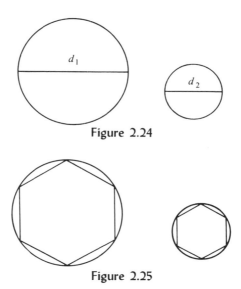

Figure 2.24

Figure 2.25

true for the areas of these polygons, it seems as though it should be true for the circles themselves. This was the type of argument probably used by Hippocrates some 50 or 60 years earlier than Eudoxus, but it was not quite convincing. There was much paradoxical and confusing discussion about whether or not a circle could be thought of as a polygon of infinitely many sides, and the question had never been answered.

Eudoxus found a way to prove his theorem avoiding this discussion. His method, which was later applied to many other problems, became known as the *method of exhaustion*. His reasoning, a proof by contradiction, went like this.

*Proof* Suppose the theorem is not true. Then either

$$\frac{A_1}{A_2} < \frac{d_1^2}{d_2^2} \quad \text{or} \quad \frac{A_1}{A_2} > \frac{d_1^2}{d_2^2}$$

Suppose the first alternative holds. Then

$$\frac{A_1}{S} = \frac{d_1^2}{d_2^2}$$

where $S$ is an area $< A_2$ (fig. 2.26). Inscribe similar polygons, say hexagons, in circles $A_1$ and $A_2$. Continue bisecting and forming new inscribed polygons until the polygon inscribed in $A_2$ has area greater than $S$. Call these similar polygons $P_1$ and $P_2$ (fig. 2.27). It is known,

New Methods and Ideas

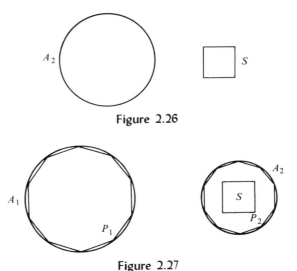

Figure 2.26

Figure 2.27

by the theorem for polygons, that

$$\frac{P_1}{P_2} = \frac{d_1^2}{d_2^2}$$

which gives

$$\frac{A_1}{S} = \frac{P_1}{P_2}$$

But $A_1 > P_1$ and $S < P_2$, thus this is impossible. The other case is done similarly.

To complete this proof Eudoxus did not require that the polygons actually increased in number of sides to infinity. In using a proof by contradiction, he only needed the polygons to increase their number of sides until $P_2 > S$. Thus, Eudoxus avoided the logical problems of a circle being a polygon of infinitely many sides.

Eudoxus applied his method of exhaustion to prove that a cone has a volume equal to one-third of the volume of the cylinder in which it is inscribed (fig. 2.28). He approximated each figure by polygonal prisms to arrive at his proof. (We will not outline the proof in this text.)

Democritus, who lived at about the same time as Hippocrates, had tried this problem of determining the volume of a cylinder, but encountered a similar paradox as did Hippocrates with his problem concerning

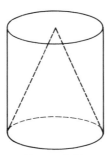

**Figure 2.28**

the areas of circles. Democritus may have assumed the cone to be made up of thin infinitesimal slices (fig. 2.29a). But if two neighboring slices are equal, the figure would be a cylinder (fig. 2.29b). If they are unequal, the figure would have steps (fig. 2.29c). Fortunately, Eudoxus found a method to avoid dividing the cone into infinitely many slices.

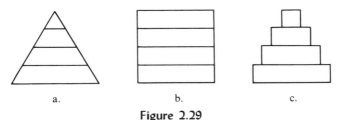

**Figure 2.29**

Similar paradoxes arose in the 1600s when problems like these were again studied, and calculus was developed. Infinite sets proved to be a difficult concept for the early mathematicians.

It was in the time of Eudoxus that the conic sections were first studied. These problems may have arisen from the shapes of shadows on a sundial, or from Indian peg and cord constructions which we studied earlier. **Menaechmus (ca. 350 B.C.)** defined the parabola as the curve obtained by slicing a right-angled cone by a plane perpendicular to a generator, the ellipse as the curve obtained by slicing an acute-angled cone by such a plane, and the hyperbola as the curve obtained from such a slice of an obtuse-angled cone (fig. 2.30).

**Aristotle (ca. 340 B.C.)** was a pupil of Plato and a teacher of Alexander the Great. Many of his works survived, so we fortunately have accurate records of his philosophy. He organized logic, and until recently his work was the standard on the subject. His theories of mechanics were studied in the Middle Ages, when his work was especially influential, and discussion and criticism of his theories were important in the development of the modern science of mechanics.

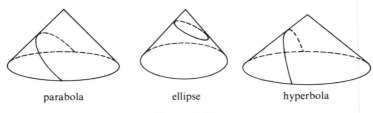

parabola  ellipse  hyperbola

**Figure 2.30**

Aristotle was more empirical in his thinking than Plato, basing many of his theories on experience. For example, he made many observations in the field of biology. Aristotle also founded a school called the Lyceum. It was here that he earned his nickname "the Peripatetic," because he paced when he taught.

## 6  The Elements — A Summary

The School of Alexandria marked a new period in the history of mathematics. Alexander the Great had conquered much of the known world, and Greeks settled in these conquered territories. In Egypt, Alexander named a city after himself, Alexandria, and a school was founded in Alexandria which was perhaps the greatest of ancient times. The school existed for about 800 years, from 300 B.C. to A.D. 500. During this period Alexandria was the center of "Greek" mathematics.

### EUCLID (ca. 300 B.C.)

Euclid, who lived in Alexandria, wrote the world's most successful textbook. His *Elements* has dominated the study of geometry ever since it was written, over one thousand editions ago; it has been second only to the Bible in popularity. Euclid did what a textbook writer typically does — he organized the material already available using his own methods where appropriate. The *Elements*, as indicated by the title, covers fundamental topics in mathematics from several areas. Euclid's work is really the first Greek mathematics (except for a few fragments) to survive. Not all his writings survived, only five out of 10, but enough were preserved to give us some records of Greek mathematics. In this text, we will limit our consideration to the most important of Euclid's works, the *Elements*.

Since Euclid compiled the *Elements* from some already existing material, it is very hard to tell how many proofs are original with him and how many were left in the form in which he found them. Euclid himself was a competent mathematician, so he could very well have developed many of the concepts in the *Elements*. Later commentators on the work gave the names of those they believed to be originally responsible for the material. In any case, from historical studies of the *Elements* we can obtain much information on the content of Greek mathematics from 550–300 B.C. Euclid's original contributions can be studied from his other more advanced works.

Euclid's book has been one of the most popular of all history, yet virtually nothing is known about Euclid himself. For a long time he was confused with another Euclid who was a philosopher. Only two stories are known about him. One concerns someone who had begun to read geometry with Euclid. When he had learned the first theorem he asked his teacher, "But what shall I get by learning these things?" Euclid called his slave and said, "Give him three cents, since he must make gain from what he learns." The only other anecdote about Euclid which has survived involves Euclid and the king. The king asked Euclid if there was not an easy way to learn geometry, and Euclid replied that "There is no royal road to geometry."

The *Elements* is divided into 13 books, similar to chapters. We will take a brief look at each book to get an idea of how the *Elements* was arranged.

Five postulates and five common notions are given at the beginning of Book I. The common notions are about equality, such as *the whole is greater than the part*. Recall that a possible reason for their inclusion in Pythagorean geometry was given in connection with the discussion of Zeno. The five postulates are more directly related to geometry than the common notions are. The first three were already discussed; they gave the construction rules, allowing points to be joined, lines to be extended, and circles to be constructed. The following statements are the last two postulates.

**Postulate I.4**   All right angles are equal to one another.

√ The fifth postulate is the famous *parallel postulate*.

**Postulate I.5**   If a straight line falling on two straight lines makes the interior angles on the same side less than two right angles, the two straight lines, if produced indefinitely, meet on that side on which the angles are less than the two right angles.

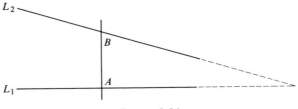

Figure 2.31

Postulate I.5 is easier to understand if reference is made to figure 2.31. The angles $A$ and $B$ are interior angles on the same side less than two right angles. Thus, Postulate I.5 asserts that lines $L_1$ and $L_2$ intersect on that side.

There is much history concerning the parallel postulate. It seemed that such a complicated statement should not be accepted without proof, so for 2200 years mathematicians tried to prove the fifth postulate. We will discuss the attempts to validate the parallel postulate in a later chapter.

Book I of the *Elements* contains 48 theorems, over half of which are used in the chain of results leading to I.47, the Pythagorean theorem. Thus, one goal of Book I may have been to give a proof of the Pythagorean theorem not depending on the false assumption of the commensurability of any two lines. Euclid uses the concepts of congruent triangles (in place of similar triangles) and parallel lines to prove Theorem I.47. These are familiar concepts of high school geometry; certain books of the *Elements* have served as models for geometry courses ever since they were written. It is interesting to me that these concepts of congruence and parallels were not aimlessly thrown together at the beginning

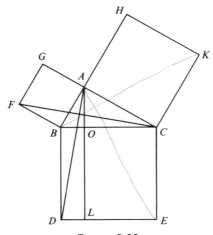

Figure 2.32

of geometry but are there because they are used in proving the Pythagorean theorem.

The following is an outline of the proof of I.47 (fig. 2.32).

*Proof*  Triangle *FBC* is proved congruent to triangle *ABD*. It is known from a previous theorem that

$$\triangle ABD = \frac{1}{2} \text{ rectangle } BOLD$$

and

$$\triangle FBC = \frac{1}{2} \text{ square } BFGA$$

Therefore,

$$\text{square } BFGA = \text{rectangle } BOLD$$

Similarly,

$$\text{square } AHKC = \text{rectangle } CELO$$

so the square on the hypotenuse *BC* is equal to the sum of the squares on the sides.

We see that this figure is more complicated than that of the Pythagoreans (see fig. 2.7), but the proof using it does not require the lines to be commensurable.

Book I contains many other theorems familiar from high school geometry, such as the following:

**Theorem I.10**  A line segment can be bisected.

**Theorem I.15**  If two straight lines cut one another, the vertical angles are equal.

The topics of Book I were probably studied by the Pythagoreans, though not much evidence from that period remains. Theorem I.4 is particularly interesting because it may reflect an older method of proof which was used long before Euclid's time.

**Theorem I.4**  If two sides and the included angle of one triangle are respectively equal to two sides and the included angle of another triangle, then the two triangles are congruent.

*Proof*  Let △*ABC* and △*DEF* be given with *AB* = *DE*, *BC* = *EF* and ∠*ABC* = ∠*DEF* (fig. 2.33). Take triangle *ABC* and place it on top of triangle *DEF*. From the given, *AB* fits exactly on *DE*, and *BC* fits on *EF*. Suppose *AC* did not fit on *DF* (fig. 2.34). Then two lines, *AC* and *DF*, would enclose a space, which is impossible.

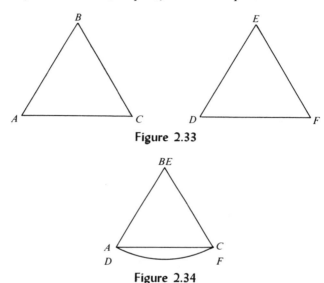

Figure 2.33

Figure 2.34

In this proof Euclid says to take one triangle and plunk it on another and look at the result. This resembles the older methods of reasoning in which visualization was important.

Two other interesting theorems are I.32 and I.29. These theorems are evidence of qualities that have made Euclid so popular — clever ideas and clear and reasonably easy proofs.

**Theorem I.32**  In any triangle, if one side is produced, the exterior angle is equal to the two interior and opposite angles (fig. 2.35).

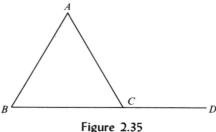

Figure 2.35

*To prove* $\angle ACD = \angle BAC + \angle ABC$

[*Note:* This theorem implies that the sum of the angles of a triangle is equal to a straight angle. Since $\angle ACD + \angle ACB$ is a straight angle, substituting we have $\angle ABC + \angle BAC + \angle ACB$ is a straight angle.]

*Proof* Construct $CE$ parallel to $AB$ (fig. 2.36). (**I.31**)
Then

$$\angle ECD = \angle ABC$$

and

$$\angle ACE = \angle BAC \qquad (\textbf{I.29})$$

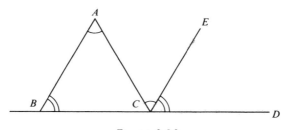

Figure 2.36

Also,

$$\angle ACD = \angle ACE + \angle ECD$$

so by substitution

$$\angle ACD = \angle BAC + \angle ABC \qquad \text{Q.E.D.}$$

I find Euclid's construction of the parallel $CE$ a very appealing solution to the problem. In addition to Euclid's proof given above, we have a report of an earlier proof of the same theorem. Rarely is such information available. **Proclus (A.D. 500)** wrote that **Eudemus (320 B.C.)** credited the Pythagoreans with proving that the sum of the angles of a triangle equals a straight angle by constructing a parallel to $BC$ through the point $A$ (fig. 2.37).

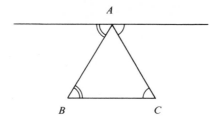

**Figure 2.37**

Theorem I.29 is also useful in the proof of I.32, and it illustrates the application of the parallel postulate. We will prove only the first part of the theorem.

**Theorem I.29** A straight line falling on parallel straight lines makes the alternate angles equal to one another.

*To prove* $\angle AGH = \angle GHD$

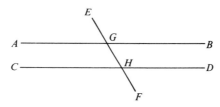

**Figure 2.38**

*Proof* Again, the proof is by contradiction. Suppose $\angle AGH \neq \angle GHD$. Let $\angle AGH > \angle GHD$. Add $\angle BGH$ to each. Then,

$$\angle AGH + \angle BGH > \angle BGH + \angle GHD$$

But the angles $AGH$ and $BGH$ are equal to two right angles (Theorem I.13), therefore, the angles $BGH$ and $GHD$ are less than two right angles. But straight lines produced indefinitely from angles less than two right angles meet (Postulate I.5). Thus, $AB$ and $CD$ will meet if produced indefinitely. But they do not meet, because they are parallel by hypothesis. Hence, $\angle AGH$ is not greater than $\angle GHD$. Similar reasoning shows that neither is it less. Therefore

$$\angle AGH = \angle GHD \qquad \text{Q.E.D.}$$

Book II of the *Elements* is comprised of material originally attributable to the Pythagoreans. It contains geometric solutions to problems that the Babylonians solved by means of their rules equivalent to the quadratic formula. When applying these rules, the Babylonians might have had to approximate a square root, as we might approximate $\sqrt{5} \cong 2.23$. Thus, the answer to the problem would be only approximate, though quite sufficient in accuracy for practical purposes. Greek mathematics was much more speculative and closer to philosophy than the practically-oriented Babylonian mathematics. Therefore, the Greeks desired the exact solution. To obtain exact answers, they remained in the domain of geometry where they could construct the precise solution to the problem, even if it was a line incommensurable with the unit. As an example, consider the following theorem.

**Theorem II.11**  A straight line can be divided so that the rectangle contained by the whole and one of the segments is equal to the square on the remaining segment.

That is, given a line $AB$ we want to cut it at $H$ so that the rectangle $HBDK$ (here $BD = AB$, the whole) = the square $AFGH$ (fig. 2.39).

In equations, letting $AB = a$ and $AH = x$, the unknown, we have the area of the rectangle $a(a - x)$ equal to $x^2$ the area of the square

$$a(a - x) = x^2$$

Theorem II.11 states and proves the following method for constructing the point $H$.

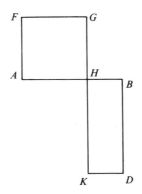

Figure 2.39

*Method* Construct a square *ABDC* on *AB*.
Bisect *AC*; call the midpoint *E*.
Connect *EB*.
Extend *AC*.
Lay off *EF* = *EB* on *AC* extended.
Construct the square *FGHA*.
*H* is the required point.

The solution *AH* is constructed (fig. 2.40). Consider an algebraic example when $a = 1$. The equation is

$$1(1 - x) = x^2$$

with solution

$$x = \frac{-1 + \sqrt{5}}{2}$$

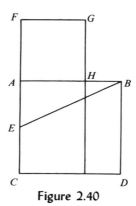

Figure 2.40

Here in a numerical solution the square root would be approximated as 2.23, for example, and an approximate value for *x* obtained. The Greeks avoided this approximation, but, as you can see, their geometric solution is rather cumbersome in comparison to the numerical rules of the Babylonians and not as useful for practical problems.

You will notice that the negative root of the equation was not mentioned in the previous calculation. Both the Babylonians and the Greeks gave only positive solutions to their problems. Actually, it took quite a long time before negative roots were accepted to equations. After all, if an answer represented a number of bricks in an area, what sense is there to $-5$ bricks? Even after negative numbers themselves were given as interpretation, negative roots were rejected as meaningless.

Book III contains theorems on circles, many of which are familiar from high school geometry courses. Presently, the theorems are the same, but the approach has been revised in keeping with modern developments. As an example from Book III, consider the following very useful theorem.

**Theorem III.31**  An angle in a semicircle is right.

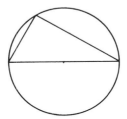

Figure 2.41

Some of the material in an earlier form of Book III may have been included in the lost *Elements* of Hippocrates. One topic studied by earlier geometers, but little mentioned by Euclid, is the angle between curved lines such as circles (fig. 2.42a). Euclid does consider in Theorem III.16 the angle between a circle and a tangent (fig. 2.42b), but usually the angles are rectilineal, that is, between two straight lines (fig. 2.42c).

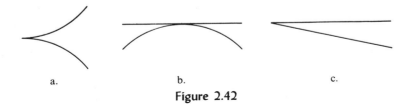

Figure 2.42

Book IV is concerned with polygons, a subject probably also considered by Hippocrates. Regular polygons of three, four, five, six, and 15 sides are inscribed in a circle, as was already described in connection with Hippocrates.

Book V contains the theory of ratios of magnitudes originally formulated by Eudoxus. His definition of one ratio being greater than another has already been discussed. An example of another useful theorem about ratios found in Book V is the following:

**Theorem V.11**  If $a:b = c:d$ and $c:d = e:f$, then $a:b = e:f$.

The theory of ratios of magnitudes is applied to geometry in Book VI. Theorem VI.4 states the most useful property of similar triangles.

**Theorem VI.4** Triangles with equal angles have their corresponding sides proportional.

To prove $\frac{AB}{DE} = \frac{AC}{DF}$, etc.

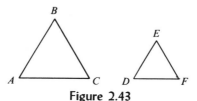

Figure 2.43

Using this theorem Euclid can perform the old Pythagorean "proof" of the Pythagorean theorem correctly (see fig. 2.7). It is given in a slightly generalized form in Theorem VI.31.

The proof is the same as the old one, except that now the corresponding sides of similar triangles were proportional in the sense of Eudoxus' theory of ratio of magnitudes. Euclid could not put this proof at the beginning of his *Elements*, because it required the more difficult concepts of the theory of ratios of magnitudes, even though it is easier to understand than the proof in I.47 which relied only on the theory of Book I. Since the Pythagorean works are lost, one can only surmise that the proof in Book VI is a corrected version of the original Pythagorean proof.[12]

Book VII was probably compiled from Pythagorean mathematics during the period 500–450 B.C. It concerns numbers and divisibility. Theorem VII.2, for example, shows how to find the greatest common divisor of two numbers (the largest number that divides both). This method is now called the *Euclidean algorithm*. As an illustration, consider the two numbers 145 and 30. The algorithm works in the following way:

$$
\begin{array}{r}4\\30\overline{)145}\\120\\\hline 25\end{array}
\qquad
\begin{array}{r}1\\25\overline{)30}\\25\\\hline 5\end{array}
\qquad
\begin{array}{r}5\\5\overline{)25}\\25\\\hline\end{array}
$$

Divide 145 by 30. Divide the remainder, 25, into the smaller of the two numbers, 30. Repeat until a remainder of zero is reached. The last

---

[12]Daniel Shanks, *Solved and Unsolved Problems in Number Theory*, p. 124.

nonzero remainder obtained is the greatest common divisor. In this case it is 5. Book VIII continues with the study of numbers, dealing with progressions, such as

$$1, p, p^2, \ldots, p^n$$

This book is probably due to Archytas and related to his studies in music.

Numbers again are the concern of Book IX. The final section, as was mentioned earlier, is the oldest part of the *Elements* and deals with even and odd numbers. Recall that Theorem XI.36 about perfect numbers is the culmination of the work with even and odd numbers.

Book X is about incommensurable lines and classifies them. It is a long and difficult book, possibly derived from the work of Theaetetus. We will not discuss it in detail, due to the complexity of the topic.

Solid geometry, a course that used to be studied in high school in Euclidean form, but is now studied analytically in college, is the subject of Book XI. This material probably originated with the Pythagoreans. Book XI contains theorems on lines and planes in three dimensions, thus the term *solid* rather than *plane* is applied to this particular form of geometry.

Book XII contains the work of Eudoxus on ratios of areas and volumes, as mentioned earlier, and Book XIII describes the construction of the five regular polyhedra inside a sphere. Euclid may have drawn from the work of Theaetetus for his discussion of polyhedra.

It is important to remember that, although Euclid's *Elements* was written about 300 B.C., the earliest manuscripts containing the Greek text date from the tenth century A.D. Thus, even the oldest available texts are copies of copies of copies, etc. It is from these works that scholars try to establish what parts Euclid himself wrote and what parts were added by later commentators and copyists. J. L. Heiberg, the Danish classical scholar, based on his extensive study of the available manuscripts, prepared as accurate a Greek text as possible of the *Elements* published in several volumes during the period 1883–88. This work formed the basis for the English translation by Sir Thomas L. Heath in which he provides an excellent introduction and much valuable commentary.

## 7 The Pinnacle of Greek Geometry

Euclid had gathered in the *Elements* much of the Greek mathematics of the several hundred years prior to his time. Greek geometers, as their techniques had become more sophisticated, tackled and solved problems of increasing difficulty. In the century following Euclid, Greek geometry culminates in several masterful works.

## ERATOSTHENES (230 B.C.)

Eratosthenes excelled in geography, mathematics, and astronomy, among other fields. He developed a method for finding prime numbers, now called the *sieve of Eratosthenes*. His method consisted of writing the integers in a list beginning with 2.

2, 3, 4̸, 5, 6̸, 7, 8̸, 9̸, 1̸0̸, 11, 1̸2̸,

13, 1̸4̸, 1̸5̸, 1̸6̸, 17, 1̸8̸, 19, 2̸0̸, 2̸1̸, 2̸2̸, 23, 2̸4̸, . . . .

The first number on the list, 2, is prime (has no integral divisors other than one and itself), but all multiples of 2 can be eliminated since they are divisible by 2. We cross out these multiples. The next number remaining on the list, 3, is prime. But, again, we can cross out all multiples of it, 6, 9, 12, 15, . . . . In this case, all the remaining numbers on the list are prime since any number less than 25 must have, if not prime, one factor less than 5. Of course, this same process can be followed with a larger list of numbers.

Eratosthenes also accurately calculated the circumference of the earth. The sun on the summer solstice was directly overhead at a city called Syene. At the same time in Alexandria the sun was inclined at an angle of 7°12′ (fig. 2.44). The distance from Syene to Alexandria was found to be 5000 stadia by someone who had walked it. Now 7°12′ = 1/50 of 360°. Substituting, 5000 stadia = 1/50 of the circumference of the earth. Therefore, the circumference is 250,000 stadia. Although we do not know the exact length of a Greek stadium, we do know that it was approximately equal to 200 yards, so Eratosthenes' result was a good estimate of the earth's circumference.

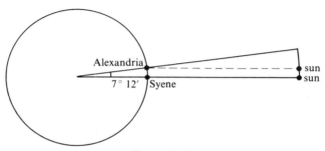

Figure 2.44

## ARCHIMEDES (225 B.C.)

Archimedes is considered to be one of the greatest mathematicians of all time. He was able to find volumes, areas, and lengths of figures which no one else could determine. Eudoxus' method of exhaustion and his own clever ideas were used in his calculations. These original methods of Archimedes involve finding limits of approximations by polygonal figures, analogous to what is now done in the calculus. Archimedes also founded the science of hydrostatics.

There are several stories about Archimedes which demonstrate his genius and eccentricities. To rescue a ship stuck in the surf Archimedes designed an apparatus with pulleys and levers so that the king (Hiero) could free the ship single-handedly. Archimedes said, "Give me a spot where I can stand, and I shall move the earth."

Archimedes had incredible concentration, which is true of many great mathematicians (Newton, for example). He could work on a problem for hours without stopping for meals. In fact, Archimedes became so engrossed in his mathematics that he would not bother with necessities such as bathing. When he was finally taken to bathe he occupied himself by drawing figures in the ashes of the fire used to heat the water. After the bath when his body was annointed with oil, he drew figures in the oil, never breaking his concentration.

One famous story about Archimedes concerns his solving the problem of King Hiero's crown while bathing. The king ordered a crown of gold but was afraid that the jeweler had mixed in some cheaper silver. He asked Archimedes to determine if the crown was of pure gold. While in the bath one day, Archimedes leapt out naked and ran down the streets shouting, "Eureka, eureka" (I have found it)! Apparently, he planned to immerse the crown in water, along with equal weights of gold and silver and measure the volumes of water displaced. Since gold is denser than silver, an amount of gold will displace less water than an equal weight of silver. If the crown is pure gold it should then displace the same amount of water as an equal weight of gold.

Another possibility is that Archimedes planned to take the crown and an equal weight of gold and weigh each under water. According to the principle of buoyancy, which he probably discovered during that same bath, a solid heavier than water will, when weighed in water, be lighter than its true weight by the weight of the fluid displaced. If the crown was impure, it would displace more water, and hence would weigh less in the water than the pure gold.

Archimedes lived in Syracuse which is on the island of Sicily, off the coast of Italy, at the time in history when the Romans were flexing their muscles, conquering new territories. When the Romans attacked Syracuse Archimedes repelled them with projectiles hurled from the walls — large rocks, fire, claws — using mechanical devices. The story is told that if the Roman soldiers saw a rope or a piece of wood extending beyond the walls of Syracuse, they ran fearing another invention of Archimedes. Eventually, however, the Romans captured Syracuse in an attack from the rear.

This information is contained in a history of the life of the Roman general who led the campaign to conquer Syracuse. The general sent a soldier to bring Archimedes to him, and the soldier found Archimedes solving a problem by tracing a diagram in the sand. Archimedes told the soldier not to bother him until he finished the problem, whereupon the soldier struck Archimedes with his sword and killed him. He was 75 at the time.

A favorite problem of Archimedes involves the ratio of the volumes of three well known solids. If figure 2.45 is revolved around the dotted line, a cone inscribed in a hemisphere, which in turn is inscribed in a cylinder, is generated. Archimedes proved that the volumes of these three figures are in the ratio 1 : 2 : 3. He was so proud of his proof that he wanted a sphere with its circumscribed cylinder and their ratio (2 : 3) engraved on his tombstone; he got his wish.

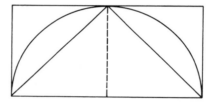

Figure 2.45

The proof of this theorem is a very elegant one. It was often wondered how Archimedes arrived at his result, and mathematicians even accused him of deliberately hiding and disguising his methods of discovery. This was not the case, however. In 1906 a lost book, *The Method*, was recovered by J. L. Heiberg in the library of a monastery in Constantinople. The text was written on parchment in a tenth century hand and had been washed off to make the precious parchment available for a book of prayers and ritual in the thirteenth century. (Such a text is called a *palimpsest* from a Greek term meaning *rescraping*.)

In *The Method* Archimedes guessed theorems by dividing solids into infinitesimally thin sections as Democritus also did. He did not believe this was rigorous enough, though, so once he discovered the theorems he found proofs for them by approximating by polygonal figures and using the method of exhaustion for the reasoning. These are problems that can now be solved by calculus, but no one reached the level of Archimedes' reasoning in these areas for over 1700 years.

Archimedes, in his description of his discovery of the 1 : 2 : 3 ratio in *The Method*, uses a slightly different figure at first (fig. 2.46) than figure 2.45. Revolving this figure about *FG*, a cone, sphere, and cylinder are generated. The line *AB* is revolved into three discs. Thus, Archimedes is considering the sphere and the other figures to be made up of thin (infinitesimal) discs. He then treats *PFG* as a lever with fulcrum at *F*. He showed that the small discs obtained from the cone and sphere, by slicing along the line *AB*, when placed at *P* just balanced the large disc of the cylinder at *AB* (fig. 2.47). From this and some further reasoning he was able to deduce the theorem.[13]

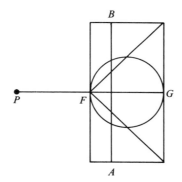

Figure 2.46

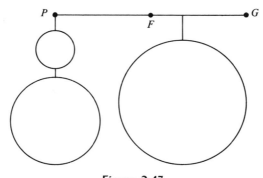

Figure 2.47

Archimedes gives what the author believes is the simplest and most elegant angle trisection method. Of course, he does not use merely the postulates of Euclid (straightedge and compass) — that has been proved

---

[13]See Asger Aaboe, *Episodes from the Early History of Mathematics*, pp. 93–98.

to be impossible — but his only additional requirement is a marked straightedge.

*Method*   Let *ASB* be the angle to be trisected (fig. 2.48).

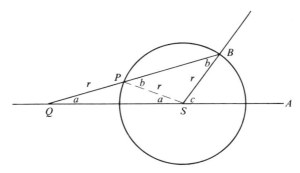

**Figure 2.48**

Construct any circle with center *S* at the vertex and some radius *r*.
Extend the line *AS*.
Mark the straightedge with a length equal to the radius. Take this marked straightedge and draw a line from *B* to *Q* on the line *AS* extended so that the distance *QP* from *Q* to the circle is equal to the radius *r*.
Then, $\angle BQA = 1/3 \ \angle ASB$.

*Proof*   The angles at the base of an isosceles triangle are equal. Thus, the angles can be labelled as indicated in the figure. $\angle BPS$ is an exterior angle of $\triangle QPS$. Thus, by *Elements* I.32,

$$b = a + a$$

Also, $\angle ASB$ is an exterior angle of $\triangle QSB$. Thus,

$$c = b + a$$

Substituting,

$$c = a + a + a$$

and

$$\angle BQS = \frac{1}{3} \angle ASB$$

Archimedes also developed an interesting method for finding an approximation for $\pi$, the ratio of the circumference of a circle to its diameter. He found that the ratio of the perimeter of an inscribed 96-sided polygon to the diameter of the circle is greater than $3\frac{10}{71}$, and the same ratio for a circumscribed 96-sided polygon to be less than $3\frac{10}{70}$. The ratio for the circle is in between the ratios of the polygons. Thus,

$$3\frac{10}{71} < \pi < 3\frac{10}{70}$$

This problem always causes trouble, because, as was shown in the 1800s, the ratio is not a whole number or fraction, but an irrational number. To obtain his approximation, Archimedes started with the side of a hexagon which he knew how to find. He then bisected the angle to get the side of a 12-gon and showed how to find the length of a 12-gon side. Archimedes used both inscribed (fig. 2.49a) and circumscribed (fig. 2.49b) polygons. He repeated the procedure, going from 12 to 24 to 48 to 96 sides. One is only limited here by the effort it takes to do the computation. Around the year 1600 several mathematicians carried this method much further.

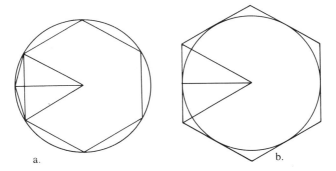

Figure 2.49

Archimedes also proved a number of theorems about *spirals*. A spiral is the curve traversed by a particle moving along a line with uniform speed, while the line rotates (fig. 2.50). (The Greeks were not familiar with very many curves. The ones they were familiar with were obtained by some simple means such as slicing a cone. It is only with algebra that it becomes easy to express millions of curves.) Archimedes proved, for example, that the area of the first turn of the spiral is one-third the area of the circle shown (fig. 2.51). His method for computing this is quite interesting, involving approximation by sectors of circles and sums like $1^2 + 2^2 + \cdots + n^2$.

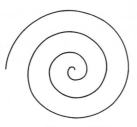

Figure 2.50           Figure 2.51

Archimedes also found the area of a segment of a parabola (fig. 2.52). He proved that the area of the segment is four-thirds of the area of the inscribed triangle with the same base. He did this by inscribing a triangle in each of the two remaining areas and showing that these triangles each have area $T/8$ where $T$ is the area of the original triangle. Then, in the four remaining areas he put four triangles each of area $T/64$, and so on. He thus approximated the area of the parabolic segment as a sum of triangles.

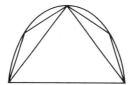

Figure 2.52

$$\text{segment} = 1 \text{ triangle} + 2 \text{ triangles} + 4 \text{ triangles} + \ldots$$
$$\text{area} = T + 2\left(\frac{T}{8}\right) + 4\left(\frac{T}{64}\right) + \ldots$$
$$= T + \frac{T}{4} + \frac{T}{4^2} + \ldots$$

He then showed that this area must be $4/3T$, as we can see by summing a geometric series.

A rather simple sounding problem which has an almost impossible solution is said to have originated with Archimedes. The numbers of white, black, yellow, and dappled cows and bulls were sought, given nine conditions, such as the number of white bulls equals $(1/2 + 1/3)$ black bulls + yellow bulls. It turns out that the least solution is so large that the number of cattle would require a number of more than 206,500 *digits*! Archimedes used to write of his discoveries to others without proof so that they could have the pleasure of discovering for themselves how to prove the problems. When some people started announcing

Archimedes' discoveries as their own, he began adding a few false statements to embarrass the thieves. He also liked to challenge other mathematicians; the cattle problem is probably one of his intellectual challenges.

We have surveyed just a sampling of the work of Archimedes. His reasoning is often complex and ingenious, and his work is typically more advanced than that of Euclid. The fact that he originated mathematical studies in the areas of levers, centers of gravity, and floating bodies is further evidence of his genius.

## APOLLONIUS (225 B.C.)

Apollonius of Perga was known as "the great geometer." His most important work was *Conics* which was composed of eight books. Only seven have survived — four in Greek and three in the Arabic translation. All his other works except one have been lost.

Menaechmus, over 100 years before, had been the first to study sections of a cone which were among the few curves known to the Greeks. Apollonius defined a double cone — two similar cones lying in opposite directions and meeting in a fixed point (fig. 2.53c). He specified that if a straight line indefinite in length and passing always through a fixed point be made to move around the circumference of a circle which is not in the same plane with the point, the moving straight line will trace out the surface of a double cone. Apollonius showed that the conic sections could all be obtained from a double cone by varying the angle of the cutting plane, no longer requiring it to be perpendicular to the generating line (fig. 2.53a & b).

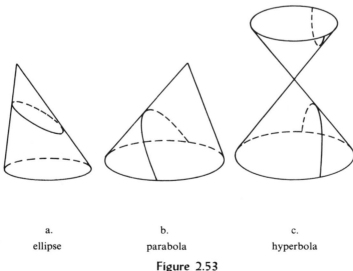

a.
ellipse

b.
parabola

c.
hyperbola

**Figure 2.53**

Apollonius gave the conic sections their current names. These names are based on a relation which resembles our equation referring to axes with the origin at a vertex of the conic section. Refer to figure 2.54. Here the diameter $d$ for the ellipse is the major axis, and for the hyperbola is the distance between the vertices of the two branches. The latus rectum, $l$, is the chord through the focus of the conic and perpendicular to the axis of the conic. (Apollonius defined the latus rectum somewhat differently.[14]) Apollonius chose the word *ellipse* to represent the fact that in the equation, $y^2$ is *less than* $lx$. Similarly, for the parabola $y^2$ *equals* $lx$, and for the hyperbola, $y^2$ *exceeds* $lx$. (Note the similar use of the words *ellipsis* and *hyperbole* in composition. An ellipsis indicates that words are left out, while hyperbole means an extravagant exaggeration of a statement.)

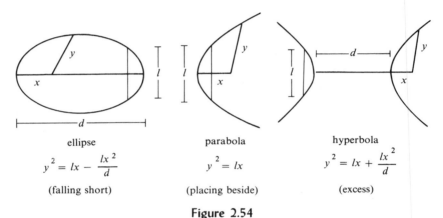

ellipse  
$y^2 = lx - \dfrac{lx^2}{d}$  
(falling short)

parabola  
$y^2 = lx$  
(placing beside)

hyperbola  
$y^2 = lx + \dfrac{lx^2}{d}$  
(excess)

**Figure 2.54**

Apollonius' work on conics contains a variety of very advanced theorems on tangents and normals. His attitude toward his work in this area is interesting — he said of the results, "They are worthy of acceptance for the sake of the demonstrations themselves, in the same way as we accept many other things in mathematics for this and for no other reason."[15]

This satisfied Apollonius as to the importance of conic sections, but 1800 years later these curves proved to be of tremendous value in the new science of motion of projectiles and in new approaches to the motion of planets. It was discovered that planets move in elliptical orbits, and that a projectile follows the path of a parabola. Conic sections have

---

[14]For Apollonius' approach, including his derivation of the relations in figure 2.54, see T. L. Heath, ed., *Treatise on Conic Sections* (New York: Barnes and Noble, 1961); or Heath, *Greek Mathematics*, pp. 347–76.

[15]Heath, *Treatise*, p. lxxiv.

become even more practical in optics and in the analysis of maximum and minimum problems in engineering. Although the conics have become very useful, they were first studied extensively because of their beauty.

The achievements of Archimedes and Apollonius mark the height of Greek advancement in geometry. In fact, they had obtained most of the results possible with the Greek techniques that were available to them. Further advances in these areas were not made until algebra had developed substantially, at which time analytic geometry and calculus were created.

# 8 The Late Period

During the period from 200 B.C. to A.D. 500 the Greeks were under the rule of the great Roman Empire. Greek culture in Alexandria still survived, educated Greeks spoke and wrote in Greek, and the Greek schools continued to exist, but during this late period there were times of political turmoil which interrupted the continuity of learning. Alexandria, where mathematicians drew not only on the earlier Greek works of Euclid and Archimedes, but also on Babylonian mathematics, remained the most important center for Greek mathematics.

Since geometry had already been studied so thoroughly, new fields rose to prominence. Mathematical astronomy, predicting the motion of the moon and planets, was advanced tremendously. To achieve this advancement it was necessary to study triangles numerically, showing how some parts of a triangle can be found if others are known. This study of triangles marks the advent of trigonometry.

**Hipparchus (140 B.C.)** wrote on astronomy, having had access to Babylonian data. Though his work is lost, some of his achievements, such as the development of a table of chords, were noted by Ptolemy. Such a table, necessary to compute the orbits of the moon and planets, was the Greek equivalent of our sine table and could be used in much the same way. Refer to figure 2.55 to see how chord $\alpha$ relates to sin $\alpha/2$. By taking half the angle

$$\sin \frac{\alpha}{2} = \frac{1/2 \text{ chord } \alpha}{R}$$

where $R$ is the radius of the circle. Hipparchus' method of construction of the chord table is unknown, but Ptolemy's will be described later.

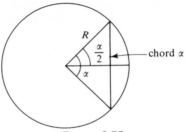
Figure 2.55

During the next 200 years, there were no mathematicians of great importance. It might help you to place this time in history to know that Julius Caesar lived during this period (100–44 B.C.).

## NICOMACHUS AND HERON

**Nicomachus (ca. 100)**[16] was from Gerasa near Jerusalem, but he lived in Alexandria. Though he did not do anything original, Nicomachus is mentioned because he represents the state of mathematics at the time. He was a Neopythagorean (*neo* means *new*), one of a sect of philosophers then flourishing in Alexandria that was trying to revive the teachings of Pythagoras.

Nicomachus wrote books on the theory of numbers and music, and these were used in the few remaining schools of philosophy. Their mathematical level was equal to that of the original Pythagoreans; in fact, Nicomachus' book is a good source for determining what the original Pythagorean subjects of study must have been like. It includes figurate numbers, perfect numbers, and number mysticism, all of which were mentioned in our earlier discussions of the Pythagoreans.

**Heron (ca. 75)** invented various mechanical devices which he regarded as toys, but he was also interested in the practical problem of finding areas and volumes of figures. Heron lived in Alexandria where he was influenced by Greek mathematics to the extent that he took results from Archimedes and gave some proofs in the Greek manner, but he also gave approximate numerical rules and formulas in the manner of the Babylonians. For example, he mentioned a formula for the area of a triangle, which does not require knowledge of the altitude,

$$A = \sqrt{s(s-a)(s-b)(s-c)}$$

where $a$, $b$, $c$ are the sides and $s = 1/2$ the perimeter. Heron gave a nice proof, but this formula may have been due originally to Archimedes.

---

[16] All dates from this point on will refer to years A.D., unless otherwise specified.

Heron also gives formulas for areas of other polygons in terms of a side. For example, the area of a pentagon is given as

$$A_5 = \frac{5}{3} s_5^2 \quad \text{and} \quad A_5 = \frac{12}{7} s_5^2$$

Neither formula is exact because an approximation to $\sqrt{5}$ is taken, but this was as good a numerical value as was needed for practical application. This, this example from Heron is similar to the numerical, approximate if necessary, approach of the Babylonians. The problem of finding a square inscribed in a given triangle (fig. 2.56), which later appeared with exactly the same numbers in the work of an Arabic mathematician, Al-Khowarizmi, was also given by Heron.

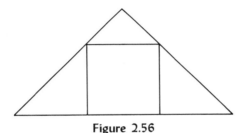

Figure 2.56

## PTOLEMY (150)

Trigonometry and astronomy were thriving at this point in history. Ptolemy was the great name in this area, and his work was the basis for astronomical theory and calculations from his day to the 1600s. His book on astronomy is the *Almagest* which means *the great work*. The name was given to it by Arabic astronomers who thought so highly of it.

Plato had felt that the perfect shapes were spheres and circles. Thus, he maintained that all motion of eternal things must be unchanging motion in perfect circles at uniform speed. When the planets are observed, however, their paths are most certainly not perfect circles traversed at uniform speed. They even appear to stop and move in the opposite direction during certain periods. Therefore, mathematicians who followed Plato, in the spirit of Greek thinking, tried to find hypotheses for the motion of the planets which would have them move in perfect circles at uniform speed and still "save the phenomena," that is, account for the appearance of the planets' motion.

In the tradition of the Pythagoreans and Plato, the mathematical explanation, the ideal form, was the reality for Ptolemy. It was not felt to be necessary to seek physical causes for the motion or even to ensure that the geometrical schemes were physically realizable. In a way, this mode of thinking is similar to current theoretical science which seeks to explain appearances. We do not take the observed planetary motion as it appears, but have a theory which explains our observations. Ptolemy also required that his theory fit his observational data.

Eudoxus and Apollonius, among others, had worked on this problem of "saving the phenomena." Ptolemy developed an epicyclic model (fig. 2.57) to represent the motion of a planet. He had the planet move on a circle, the *epicycle*, which itself rotated about another circle, the *deferent*. By tracing the path of a particle as seen by an observer, one can get quite a few possibilities for the path, depending on the size of the circles. In this model the observer was a little off the center of the deferent. The epicycle moved uniformly with respect to the equant point which was also off the center of the deferent. Actually, this system worked quite well in accounting for the motion of the planets, even though Ptolemy considered the earth fixed and the planets to be revolving about it. It is ironic that when his system was finally overthrown it was by a believer in circles and uniform motion, Copernicus, who was just trying to perfect the Ptolemaic system.

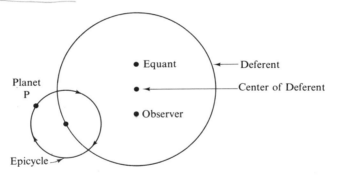

Figure 2.57

Ptolemy showed the influence of Greek mathematics strongly in his geometrical theory of epicycles. His detailed numerical computations show the influence of the Babylonians who had a purely numerical astronomy. This Babylonian influence is seen rather clearly in Ptolemy's use of sexagesimal fractions. He used the Greek alphabetic numerals and mixed systems, just as we do, in writing 112°10′13″. In a pure sexagesimal system this would be written 1,52;10,13. Of course, the base 60 system is only more convenient for writing fractions, and it is only for fractions that Ptolemy used it. Ptolemy's theory was an early example of a precise mathematical science which blends a geometrical model with numerical computation. His combining of Greek geometry and Babylonian computation has been viewed as a major step along the path to modern Western mathematical science.[17]

Ptolemy extended the trigonometry of his predecessors for use in his computations. He calculated a table of chords which was his equivalent to our table of sines. This table was much needed in astronomy and it, or one like it, served the field until modern times. It is interesting to see how he was able to construct such a table.

Recall the chord function (fig. 2.55). Ptolemy used a circle of radius 60, so

$$\sin \frac{\alpha}{2} = \frac{\frac{\text{chord } \alpha}{2}}{60} = \frac{\text{chord } \alpha}{120}$$

or

$$\text{chord } \alpha = 120 \sin \frac{\alpha}{2}$$

He constructed a table of chords in $1/2°$ steps from $1/2°$ to $180°$; thus, he tabulated chord $1/2°$, chord $1°$, $1\frac{1}{2}°$, $2°$, . . . $180°$. Ptolemy used the following method to calculate the chords.

Chord $72°$ was found by using his construction of a regular pentagon.

Chord $60°$ was known to be the radius of the circle.

Chord $12°$ was found by using the difference formula for chord $(\alpha - \beta)$ applied to the known chords of $60°$ and $72°$.

Chord $6°$ was found by using the half-angle formula for chord $(\alpha/2)$ applied to $12°$.

Chord $3°$ was found by using the half-angle formula applied to $6°$.

Chord $1\frac{1}{2}°$ was found by using the half-angle formula applied to $3°$.

Chord $3/4°$ was found by using the half-angle formula applied to $1\frac{1}{2}°$.

Chord $1°$ was found by interpolating between the values of chord $3/4°$ and chord $1\frac{1}{2}°$ and proving that the value obtained is correct.[18]

Chord $1/2°$ was found by using the half-angle formula applied to $1°$.

---

[17] See Derek Price, *Science Since Babylon*, chap. 1.
[18] Aaboe, *Episodes*, pp. 122-25.

Then the remaining chords were found by using the sum formula for chord $(\alpha + \beta)$.

We see that Ptolemy had the equivalent of our trigonometric formulas for sin $(\alpha - \beta)$, sin $(\alpha + \beta)$ and sin $\alpha/2$. He calculated them by using the following theorem.

**Ptolemy's Theorem**   Given any quadrilateral inscribed in a circle, the product of the diagonals equals the sum of the products of the opposite sides (fig. 2.58).

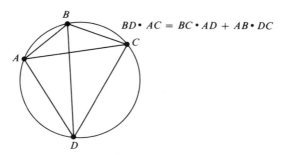

Figure 2.58

Let us see how Ptolemy used this theorem to obtain a difference formula. Suppose $AC$ = chord $\alpha$ and $AB$ = chord $\beta$ are given numerically, and we wish to find $BC$ = chord $(\alpha - \beta)$ (fig. 2.59a). [*Note:* The angles $\alpha$ and $\beta$ are not shown because this would obscure the derivation.] Let $AD$ go through the center (fig. 2.59b) so that, being a diameter, its length is known as 120 (recall that Ptolemy used a circle of radius 60). Since the angle $ACD$ is inscribed in a semicircle, it is right. Since $AC$ is given and $AD$ is known, we can find $CD$ by using the Pythagorean theorem. Similarly, $ABD$ is a right angle, and we can find $BD$. Thus, all five lines shown are known. If we draw in $BC$, we can find it by solving for $BC$ in the equation of Ptolemy's theorem in which everything but $BC$

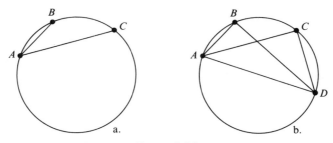

Figure 2.59

is now known. There are somewhat similar derivations for chord $(\alpha + \beta)$ and chord $(\alpha/2)$.[19]

Ptolemy needed to know the chord of 72° in order to begin the process. Recall his construction of the side of a regular pentagon (fig. 2.16). He needed to find $BF$. Since $DB$ is a radius and $DE$ is 1/2 a radius, he found $BE$ by using the Pythagorean theorem. Since $EF = EB$, he found $DF$ by subtraction, $DF = EF - DE$. Then he found $BF$ by using the Pythagorean theorem. Numerically, $DB = 60$, $DE = 30$, thus

$$BE = \sqrt{30^2 + 60^2} = \sqrt{4500}$$
$$= 67;4,55 \quad \text{(approximately)}$$

Now

$$DF = EF - DE$$
$$= 67;4,55 - 30 = 37;4,55$$

and

$$BF = \sqrt{DF^2 + DB^2} = \sqrt{4975;4,15}$$
$$= 70;32,3$$
$$= \text{chord } 72°$$

Using this value and the formulas he had developed, Ptolemy completed his table.

Ptolemy made long-lasting contributions in other areas, also. He wrote a very important work on geography from which maps were taken for approximately 1500 years. Astrology was becoming very popular in Hellenistic Alexandria at this time, and Ptolemy wrote the *Tetrabiblos*, a celebrated work in that field, which is apparently still used. He probably found his knowledge of astronomy helpful in the writing of this book. Ptolemy's works were influential for centuries, some even are today, but unfortunately, almost nothing is known about his life.

## DIOPHANTUS (ca. 250)

Diophantus of Alexandria influenced the development of algebra, particularly around 1600 when his work was rediscovered in Europe. It is not known exactly when Diophantus lived; indeed, many of the dates

---

[19]Ibid., pp. 117–20.

of this period are uncertain. We see in Diophantus' work a mixture of the Greek and the Babylonian traditions. He wrote a book, *Arithmetic*, in 13 volumes of which all but six are lost. This book concerns the theory of calculation with numbers, and as such it represents a facet of Greek studies in numbers not found in any other extant Greek work. It contrasts with other studies about the nature of the numbers themselves which consider even and odd and perfect numbers. Diophantus considered not the nature of the numbers themselves, but ways in which to divide them into parts, for example. Diophantus was only concerned with integral or fractional solutions, a necessary limitation in the Greek approach to numbers. This limitation made his problems much more difficult than modern algebra problems which appear similar. Problems requiring integral solutions are now called Diophantine, in respect to the skillful mathematician.

The *Arithmetic* contains about 150 problems, each solved by a special method. Diophantus' methods are extremely clever, as illustrated by problem 9 of Book II.

**Problem II.9** Express a given number which is the sum of two squares as the sum of two other squares.

*Solution* Let the given number be 13, the sum of the squares of 2 and 3. Let the sides of the two other squares be $s + 2$ (the 2 is chosen to match the given number 2), and $2s - 3$ (the 3 is chosen to match the given 3, while the 2 is arbitrary). Thirteen is to be the sum of these two squares also, so

$$13 = (s + 2)^2 + (2s - 3)^2$$
$$13 = 5s^2 - 8s + 13$$
$$5s^2 = 8s$$
$$s = \frac{8}{5}$$

Notice how Diophantus cleverly chose the unknowns to obtain an easy equation to solve and to insure a rational solution. Substituting for $s$, the squares are

$$(s + 2)^2 = \frac{324}{25}$$

and

$$(2s - 3)^2 = \frac{1}{25}$$

Their sum is 13.

Diophantus illustrated his method of solution by giving a numerical example. There were many solutions to this problem, but Diophantus did not specifically mention this. Other answers can be obtained by making a different arbitrary choice for the coefficient of $s$ in the second unknown, for example, choosing $4s - 3$ instead of $2s - 3$ as the side of a square.

Some rules of algebra were known to Diophantus. He mentioned transposing and used it in the course of his solutions. In giving numerical problems and using rules his work was similar to the Babylonians. To find two numbers whose sum and product were given, his approach resembled that of the Babylonians, however, he required that the square of half the sum exceed the product by a square number so that the solutions would be rational numbers.

Diophantus solved some amazing problems. For example, in problem III.19 he found four numbers such that the square of their sum plus or minus any one singly gave a square. His solutions were

$$\frac{17{,}136{,}600}{163{,}021{,}824}, \quad \frac{12{,}675{,}000}{163{,}021{,}824}, \quad \frac{15{,}615{,}600}{163{,}021{,}824}, \quad \frac{8{,}517{,}600}{163{,}021{,}824}$$

Diophantus developed a method for abbreviating frequently occurring terms. This method will be illustrated with English words rather than Greek. For the unknown number Diophantus would write *NU*, for the square of the unknown $S^q$, and for the cube $C^u$. He carried his method to the sixth power, abbreviating the fourth power as $S^qS$, the fifth as $SC^u$, and the sixth as $C^uC$. For units he wrote $U$, and for less (minus) he used *LE*. With this system of abbreviations, to write $x^3 - 5x^2 + 8x - 1$ Diophantus would first have grouped the positive terms, giving $x^3 + 8x - (5x^2 + 1)$. He would have expressed this as

$$C^u\ 1\ NU\ 8\ LE\ S^q\ 5\ U\ 1$$

## MATHEMATICS TO THE SIXTH CENTURY

**Pappus (320)** of Alexandria wrote commentaries on works of Euclid, Archimedes, and Apollonius, from which we today derive much information that would otherwise have been lost. He explained the difficult points, supplemented the texts, and added many results of his own. His biggest contribution was in keeping the earlier works alive.

**Theon (365)** was also a commentator on earlier works. He explained how Ptolemy found square roots, using the example $\sqrt{4500} = 67;4,55$. The following illustration uses decimals to show Ptolemy's method for finding the square root of two, and we will see that it gives a geometric explanation for a modern square root algorithm.

$$
\begin{array}{r}
1.\ 4\ 1 \\
\sqrt{2.00\ 00\ 00} \\
1 \\
\hline
24 \quad\ \ 1\ 00 \\
96 \\
\hline
281 \quad\ \ 4\ 00 \\
2\ 81 \\
\hline
1\ 19\ \ldots.
\end{array}
$$

Refer to figure 2.60 for Theon's explanation.

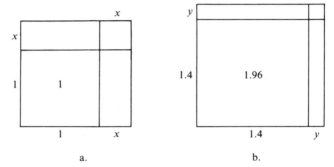

Figure 2.60

*Method* Find the largest integral square $\leq 2$. This is a square of side 1 (fig. 2.60a). Extend this to a larger square by adding a length $x$. The exact area added is $1 \cdot x + 1 \cdot x + x^2$, or $2(1)x + x^2$, which has to be less than the area of 1 needed to make a square of area 2. Thus

$$2x + x^2 \leq 1$$

We do not want to solve this quadratic inequality, so neglect the $x^2$ and get an approximation, $2x < 1$. The value $x = .5$ is too large so we choose $x = .4$. Thus, the added area is

$$x(2 + x) = .4(2.4) = .96$$

and our new square has area 1.96. We have 1.4 as an

approximation to the side of a square of area 2.

We can repeat this process as many times as we want in order to achieve greater accuracy. To begin the next iteration, extend the square of side 1.4 to a larger square of side $1.4 + y$ (fig. 2.60b). The added area is $2(1.4)y + y^2$. It must be less than or equal to $2 - 1.96 = .04$. To get an approximation neglect the $y^2$ and set $2.8y < .04$, obtaining $y \approx .01$ (to 1 significant figure). This gives an added area of $.028 + .0001$, or $.0281$. Thus, the new approximation of a square of side $1.4 + .01 = 1.41$ is less than 2 in area by $.0400 - .0281 = .0119$.

Notice the same numbers appearing in this explanation and in the usual square root algorithm. We could continue either method any number of steps further if we needed more accuracy. This algorithm is shorter than the divide and average method but much more difficult to remember.

**Hypatia (400)** was the daughter of Theon. Unusually well educated for a woman of her time, she wrote commentaries on the work of Diophantus and Apollonius. She taught Platonic philosophy in Alexandria and was a person of some influence in that city. Unfortunately, she suffered a violent death, killed by a mob who disagreed with her religious beliefs. Bickering among religious groups caused people to move away from Alexandria, many to Syria and other Middle Eastern countries. This was one way in which Greek learning began influencing Indian and Arabic culture.

Another Alexandrian, **Proclus (450)**, went to head the Platonic Academy in Athens. As do other works written at that time, his commentary on the first book of Euclid's *Elements* provides much information on early Greek geometry. Proclus having available to him sources which have since been lost.

**Boethius (500)** wrote on each of the four areas of study composing what is known as the *quadrivium* — arithmetic, geometry, music, and astronomy — but his work was very elementary. For example, his arithmetic contained the Pythagorean ideas from Nicomachus, and his geometry contained some very elementary theorems from Euclid given without proof. The books of Boethius were widely used in the monastic schools in the early Middle Ages in Europe.

In 529 the Emperor Justinian closed the school of Athens, ending Greek culture in the west. In the same year St. Benedict founded a monastery. Europe was in the process of changing from Roman ways to a culture of its own, as we will study later. For the next 500 years at least, other areas of the world are more important in mathematics.

## Problems

1. The first three perfect numbers given by the formula $2^{n-1}(2^n - 1)$ are 6, 28, and 496. Find the fourth.

2. The first three pentagonal numbers are 1, 5, and 12 (fig. 2.61). Find the next one and draw that pentagon.

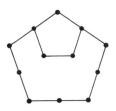

Figure 2.61

3. Write the following numbers in the Attic Greek system.
   a. 72  b. 488  c. 10,765

4. Write the following numbers in the Ionic Greek system.
   a. 27  b. 540  c. 3469  d. 78,184

5. Prove that the construction for the bisection of a line segment given in the text (fig. 2.12) does in fact bisect the line segment.

6. Let a line of length 1 be divided into two parts according to the *golden section*. Compute the length of each part correct to two decimal places.

7. Use the Euclidean algorithm to find the greatest common divisor of the following numbers.
   a. 350 and 135  b. 87 and 21  c. 317 and 249

8. Using a straightedge and compass and the method of the *Elements* II.11, construct the solution to $3(3 - x) = x^2$.

9. Use the *sieve of Eratosthenes* to make a table of the primes from 2 to 100.

10. Use Heron's formula to find the area of a triangle with sides 3, 6, and 7.

11. Show that the point $H$ obtained in the construction of Theorem II.11 of the *Elements* divides the line $AB$ according to the proportions of the *golden section*.

12. Compute the following square roots to two decimal places using the geometric explanation of Ptolemy's method given by Theon.
    a. $\sqrt{6}$      b. $\sqrt{7}$      c. $\sqrt{20}$

13. Write the following expressions the way Diophantus would have, using the symbols given in the text.
    a. $9x^3 - 3x^2 - 2x + 5$      b. $-9x^2 + 3x + 2$

14. Let $AB = x$ be the side of a regular decagon inscribed in a circle of radius 1 (fig. 2.62). Let $AC$ be constructed equal to $AB$.

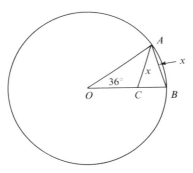

Figure 2.62

a. Show that $BC = 1 - x$.
b. Show that $\triangle OAB$ is similar to $\triangle ABC$.
c. Show that $x$ satisfies the equation $x^2 + x - 1 = 0$, which has the positive root $(1/2)(\sqrt{5} - 1)$.

15. Consider Ptolemy's construction of the regular pentagon given in figure 2.16. Assume that the radius is 1.
    a. Show that $DF$ is equal to $(1/2)(\sqrt{5} - 1)$ and is, thus, the side of a regular decagon (see problem 14).
    b. Euclid's Theorem XIII.10 states that the sides of the regular pentagon, hexagon, and decagon inscribed in a given circle form a right triangle. Use this to show that $BF$ is the side of a regular pentagon.

16. The definition of two ratios being equal given in Book V of Euclid's *Elements* states that $R/S = V/W$ if, and only if, given any integers $m$ and $n$,
    i. if $mR > nS$, then $mV > nW$.
    ii. if $mR = nS$, then $mV = nW$.
    iii. if $mR < nS$, then $mV < nW$.
    Use this definition to prove Theorem V.11 of the *Elements*.

17. Use the definition of equality of ratios stated in problem 16 and the definition of one ratio being greater than another to prove Theorem V.13 of the *Elements* which states that if $a/b = c/d$ and $c/d > e/f$, then $a/b > e/f$.

18. Prove that $A_1/A_2 > d_1^2/d_2^2$ is impossible, thus completing the proof of Theorem XII.2 of the *Elements*. [*Hint:* Let $S_2$ be an area such that $A_1/S_2 = d_1^2/d_2^2$. Use the fact that there is an area $S$, such that $S_2/A_1 = A_2/S$. The proof is then similar to the case $A_1/A_2 < d_1^2/d_2^2$ treated in section 5.]

19. Letting $AB$ in figure 2.40 equal $a$, prove algebraically that the area of the square $AFGH$ equals the area of the rectangle $HBD$, as Theorem II.11 states.

20. Prove Theorem III.31 of the *Elements* which states that an angle in a semicircle is right. [*Hint:* Letting $E$ be the center of the circle, connect $AE$. Also extend $BA$ to some point $F$ (fig. 2.63).]

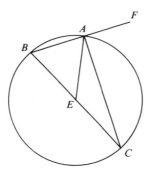

Figure 2.63

21. Explain how figure 2.1c can be used to give a proof of the Pythagorean theorem for the case of an isosceles right triangle.

22. Explain how a 15-sided regular polygon can be inscribed in a circle, given that it is already known how to inscribe both a 3-sided and a 5-sided regular polygon.

23. Let a triangle $COA$ be given and let $DO$ bisect the angle $COA$ (fig. 2.64). Theorem VI.3 of the *Elements* states that $CO/OA = CD/DA$. Use this to prove that $(CO + OA)/CA = OA/DA$. [*Hint:* Work backwards, using the relation $CA = CD + DA$.]

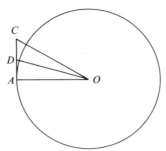

Figure 2.64

24. Suppose that angle $COA$ in figure 2.64 is $30°$. Then $CA$ is half the side of a circumscribed hexagon. Also, $CA/OC = \sin 30° = 1/2$.
   a. Show that $OA/AC = \sqrt{3}$.
   b. Given that $\sqrt{3} > 265/153$, compute an upper bound for the ratio of the perimeter of the circumscribed hexagon to the diameter. [*Note:* This also gives an upper bound for $\pi$.]

25. Use the formula of problem 23 and the values for $OA$, $CO$ and $CA$ from problem 24 to compute the ratio $DA/OA$. From this ratio compute an upper bound for the ratio of the perimeter of the 12-sided circumscribed polygon to the diameter (which is, in turn, an upper bound for $\pi$).

26. Explain how Archimedes, who used the method outlined in problems 23–25, was able to find an upper bound for the ratio of the perimeter of a 24-sided circumscribed polygon to the diameter (also the ratio of a 48-sided and a 96-sided polygon). It was in this way that he finally obtained his upper bound of $3\tfrac{10}{70}$ for $\pi$.

27. We can divide a circle of radius $R$ and its enclosed spiral into $n$ parts (fig. 2.65). Consider the circular sectors around each section of the spiral.

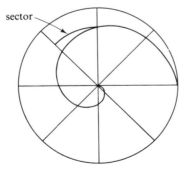

Figure 2.65

a. Use the definition of the spiral to explain why the radius of the first circular sector is equal to $R/n$, the second to $2R/n$, the third to $3R/n$, and the $n$th to $nR/n$.

b. Show that the sum of the areas of the $n$ sectors is

$$\frac{(1^2 + 2^2 + 3^2 + \cdots + n^2)\pi R^2}{n^3}$$

c. Show that the ratio of the sum of the area of the sectors to the area of the boundary circle is

$$\frac{n(n+1)(2n+1)}{(6n^3)}$$

[*Hint:* Use $1^2 + 2^2 + 3^2 + \cdots + n^2 = n(n+1)(2n+1)/6$.]

d. Show that as $n$ gets larger, the expression in part c approaches $1/3$ in value. As $n$ gets larger, the sectors become closer to the spiral. Thus, the value $1/3$ represents the ratio of the area of the spiral to its bounding circle. This, in essence, was the method of Archimedes.

28. Given an isosceles triangle of base 12 and sides 10 each, find the length of the side of a square inscribed in it (see fig. 2.56).

29. Use Ptolemy's formula for the chord of the difference of two chords to find the chord of 12°. Use the decimal system.

30. Ptolemy also used a model (fig. 2.66) in which the sun revolved on a circle whose center ($F$) was off the center of the earth ($E$). He used

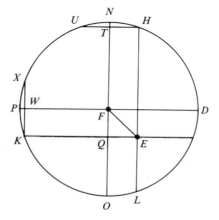

Figure 2.66

the fact that spring is the longest season, 94½ days, and summer the second longest, 92½ days, to find the distance from the center of the earth to the center of the sun's circle. Moving counterclockwise, let *NPOD* be the sun's path, with *HK* the spring portion, *KL* the summer portion, etc. Since spring is 94½ days, HK = 93°9′. Since summer is 92½ days, *KL* = 91°11′.

   a. Explain the following values.
      1) arc *HKL* = 184°20′    2) arc *NH* + arc *LO* = 4°20′
      3) arc *HN* = 2°10′        4) arc *HNU* = 2 arc *HN* = 4°20′
      5) arc *PK* = 0°59′        6) arc *KPX* = 1°58′
   b. Using Ptolemy's table of chords, it is found that *HTU* = chord *HNU* = 4;32 and *KWX* = chord *KPX* = 2;4. Find *EQ* and *FQ*.
   c. Use part b to show that the value of *EF* is between 2;29 and 2;30. (Since the radius of the sun's path is 60, it is almost 24 times greater than the distance between its center *F* and the earth *E*. Ptolemy also goes on to find that the angle *FEQ* = 24°30′.)

31. Derive Ptolemy's formula for the chord of the sum of two angles (fig. 2.67). Let *AB* and *BC* be given. The length *AC* is to be found.

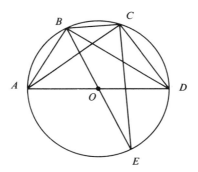

Figure 2.67

   a. The center of the circle is *O*. Explain how to find the lengths of each of the lines *AD*, *BE*, *BD*, *CE*, and *DE*.
   b. Explain how to find *CD* and *AC*, using the results of part a.

32. Show that $\sqrt{4500}$ = 67;4,55. Do the computation always using the sexagesimal system for fractions and use Ptolemy's geometric method.

33. For the regular pentagon (fig. 2.68) Heron proved that $PO + ON = \sqrt{5}(ON)$. Use this to show that the area of the pentagon is approximately $(5/3)(PQ)^2$. Use 9/4 as an approximation to $\sqrt{5}$.

Problems   91

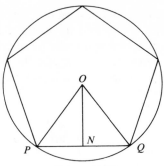

Figure 2.68

34. a. Show that there is a solution to Diophantus' problem II.9 (page 82) for each positive integer $k$, when the unknown numbers are taken as $ks - 3$ and $s + 2$.
  b. Compute the solution for $k = 3$ and for $k = 4$.
  c. Show that $k = 2$ is the smallest positive integer for which both unknowns are positive.
  d. Show that if Diophantus had chosen to let his unknowns be $ks + 3$ and $s + 2$, then he never would have obtained any positive solutions.
  e. Show that had Diophantus chosen his unknowns as $ks - 2$ and $s + 3$, then $k = 4$ would be the smallest positive integer which gives positive solutions.

35. a. The number 20 can be written as $2^2 + 4^2$. Use the method of Diophantus to write it as the sum of two other squares. Negative solutions at any stage are not allowed.
  b. Do the same as part a for $25 = 3^2 + 4^2$. Obtain a solution other than $0^2 + 5^2$.

36. In Problem III.6 of his *Arithmetic*, Diophantus finds three numbers such that their sum is a square and the sum of any pair is a square. In effect, he finds numbers $w$, $y$, $z$, $a$, $b$, $c$, and $d$ such that

$$w + y + z = a^2$$
$$w + y = b^2$$
$$y + z = c^2$$
$$w + z = d^2$$

  a. Following Diophantus, let $a = x + 1$, $b = x$, and $c = x - 1$. Then show that $w + z = 6x + 1$.

b. Choose $d$ to be any number, say 11. Then find $x$, and from it find $w$, $y$, and $z$.
c. Find a solution different from that found in part b.

37. a. List all divisors of $2^4(2^5 - 1)$, showing that each is either of the form $2^k$ where $0 \leq k \leq 4$ or of the form $2^k(2^5 - 1)$ where $0 \leq k \leq 4$.
b. Use part a to show that 496 is a perfect number.

38. Assume $2^n - 1$ is a prime number.
a. List all divisors of $2^{n-1}(2^n - 1)$ of the form $2^k$ where $0 \leq k \leq n - 1$ or of the form $2^k(2^n - 1)$ where $0 \leq k \leq n - 2$.
b. Assuming that the divisors listed in part a represent all divisors of $2^{n-1}(2^n - 1)$ except itself, show that $2^{n-1}(2^n - 1)$ is a perfect number. [*Hint:* Use the formula $1 + 2 + 2^2 + 2^3 + \cdots + 2^s = 2^{s+1} - 1$ for any $s$.]


## References

Introduction

Burnet  
Perelman  
Robinson  
Sambursky

The Beginnings of Greek Mathematics

Euclid (Heath ed.)  
Szabo (1,2)  
van der Waerden

Crises and the Origin of Deductive Mathematics

Russell  
Shanks  
Szabo (1,2)  
van der Waerden

Greek Number Systems

Boyer (1)  
Heath

Early Results and Problems of an Independent Mathematics

Heath  
Seidenberg (2,3)  
van der Waerden

New Methods and Ideas

Euclid (Heath ed.)  
Lasserre  
Neugebauer (2)  
Russell  
Seidenberg (3)  
van der Waerden

*The Elements* — A Summary

    Aaboe                             Shanks
    Euclid (Heath ed.)           van der Waerden
    Heath

The Pinnacle of Greek Geometry

    Aaboe                             Boyer (1)
    Apollonius (Heath ed.)       van der Waerden
    Archimedes (Heath ed.)

The Late Period

    Aaboe                             Neugebauer (1)
    Boyer (1)                        Price
    Diophantus (Heath ed.)       Ptolemy
    Hanson                          Sarton
    Heath                             Thomas
    Klein                              van der Waerden

# 3 Mathematics in Asia

The leading centers of mathematics in the period 500–1300 were in Asia. During that time, methods of computation with Hindu-Arabic numerals, as well as elementary techniques of algebra, were developed. Indian and Arabic writers built upon the foundation of Babylonian computational mathematics. In turn, their works were studied by European writers who learned arithmetic and algebra using translations of Arabic books. In addition, the Greek classics such as Euclid's *Elements* and Ptolemy's *Almagest* were first transmitted to European scholars in Arabic versions. Before beginning to survey the mathematics of China, India, and the Arabic world, we will study the abacus.

## I  The Abacus

There are two very simple reasons for the use of the abacus: the various types of numerals in use in ancient times were not suited to computation, and cheap paper was not available. To do ordinary calculations on papyrus or parchment made from sheepskins was too expensive. It was not until the 1300s that a rag paper was developed, and wood pulp paper did not come into use until the 1800s. Thus, when calculation was needed, the abacus was used.

The word *abacus* comes from a Greek word *abax*, meaning slab, which is related to the Hebrew word *abhaq*, meaning dust. The oldest type of abacus was a table covered with sand or fine dust, on which the figures were drawn with a stick and erased when necessary. Other forms of the abacus have permanent rows and counters on the rows to indicate numbers. Most often the counters increase in value by multiples of ten from one row to the next. We will consider five different types of abacuses.

1. The Roman abacus was a metal plate with grooves. Counters were fixed in these grooves so that they could slide up and down.
2. The line abacus, used in Roman times and in Europe until the 1600s and later in Germanic countries, was a table with lines scratched on it. Counters were placed on these lines, but were not attached to the table.
3. The Chinese abacus (Suan-pan, fig. 3.1) was not used until about 1200. (Before this the Chinese used bamboo rods for calculations, a process which will be described later.) The Chinese abacus has rods which pass through the counters enabling the counters to slide along the rods. Every rod has two counters each representing five units and five counters each representing one unit.

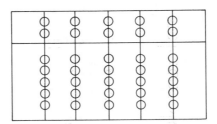

Figure 3.1

4. The Japanese abacus (Soroban, fig. 3.2) is similar to the Chinese, except for the counters. Each row of the Soroban

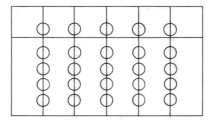

Figure 3.2

has only one counter representing five units and four counters representing single units.

✓ 5. The Russian abacus (Shchot) is also similar to the Chinese, but it has ten counters on each rod, each counter representing one unit. The middle two counters may be a different color to aid in calculating.

The abacus is still used in many parts of the world. A good abacist can calculate with the same efficiency, if not more, as a good adding machine operator. To be an expert with an abacus, one must learn to use it as a child and practice to develop skills. In Japan there are tests that are given to certify a level of competency on the abacus. One obvious advantage of the abacus is that it is certainly cheaper than an adding machine.

In the course of a calculation on an abacus, if a row becomes full, say with ten counters, then these can be replaced by one counter in the next higher row (fig. 3.3). A counter is "carried" to the next column. This is the origin of our use of "to carry" in arithmetic. In Rome the name for the counters was *calculi*. *Calculus* means *pebble* and is derived from the root word *calx*, a piece of limestone, or chalk, which comes from the same root. From the same root also comes *calculate*.

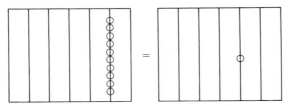

Figure 3.3

Addition on the abacus is easy. We merely enter each number in succession and read off the final answer. Multiplication can be done by successive addition, not requiring knowledge of multiplication tables. For example, 34 × 53 could be found by entering four 53s, one after another, using the first two columns, and then entering three 53s, one after another, using the second and third columns. Shifting the starting column from the first to the second is the equivalent of the familiar shifting of the products one place to the left in the usual written multiplication method. Although multiplication on the abacus can be done as was just suggested, it is actually done using a knowledge of multiplication tables. The process is similar to written multiplication, except that the intermediate results are entered on the abacus instead of being written down. For example, the same problem, 34 × 53, is done by entering 4 × 3 in the first two columns, 4 × 5 and 3 × 3 in columns two and three, and 3 × 5 in columns three and four.

The abacus may have influenced the use of the decimal system. In recording calculations done on the abacus, numerals were used. Only numerals from 1 to 9 were needed, because one could express a result as 6 hundreds, 5 tens, 3 units. This could have been abbreviated to 653, although this sort of abbreviation requires a zero to avoid confusion. For example, a result of 6 hundreds and 3 units must be written as 603, not 63.

The line abacus was used extensively in the Middle Ages in Europe at the same time that Hindu numerals were being used for calculation on a dust board or paper. There was competition between the two systems. The use of the numerals eventually triumphed, although the line abacus was in common use until about 1600. The line abacus prevailed even longer in some regions.

## 2 Chinese Mathematics

The dates of mathematical writings in China are very uncertain, but we do know that one of the oldest topics was magic squares. It is reputed that the Emperor Yu (ca. 2200 B.C.) saw such a square (fig. 3.4) on the back of a turtle, of course in a somewhat different notation. A $3 \times 3$ magic square contains the first nine integers arranged so that all rows add up to fifteen, as do all columns and the two diagonals. Fortune-tellers in the Orient still use magic squares, and they have been used as a charm in India.

| 4 | 9 | 2 |
|---|---|---|
| 3 | 5 | 7 |
| 8 | 1 | 6 |

Figure 3.4

Fifteen is the only sum possible for any $3 \times 3$ magic square using the first nine integers. We can see this by summing all the numbers in the magic square which gives $1 + 2 + 3 + 4 + 5 + 6 + 7 + 8 + 9 = 45$. These nine numbers are distributed in three rows in such a way that each row has the same sum. Thus, the sum for each row is $45/3 = 15$.

We can, in a similar manner, figure out what the sum of any row, column, or diagonal must be for an $n \times n$ magic square. The first $n^2$ numbers are distributed among $n$ rows giving a sum per row of

$$\frac{1 + 2 + \cdots + n^2}{n} = \frac{n^2(n^2 + 1)}{2n}$$
$$= \frac{n(n^2 + 1)}{2}$$

For $n = 4$ this sum is 34.

## THE NINE CHAPTERS

Probably the most influential Chinese mathematics book, and one of the older works, is the *Nine Chapters on the Mathematical Art* written about 250 B.C. This work is somewhat like Babylonian and Egyptian mathematics in that many numerical problems of a practical nature are solved. The results are sometimes exact and sometimes approximate. Rules are given for finding areas.

One type of problem found in the *Nine Chapters* much earlier than anywhere else is simultaneous linear equations. The first problem of chapter VIII is an example of this type.

> Three sheafs of good crop, two sheafs of mediocre crop, and one sheaf of bad crop are sold for 39 dou. Two sheafs of good, three mediocre, and one bad are sold for 34 dou; and one good, two mediocre, and three bad are sold for 26 dou.
> What is the price received for a sheaf of each of good crop, mediocre, and bad crop? The answer is $9\frac{1}{4}$ dou for the good, $4\frac{1}{4}$ for the mediocre, and $2\frac{3}{4}$ for the bad.

We would express this problem in equations as

$$3x + 2y + z = 39$$
$$2x + 3y + z = 34$$
$$x + 2y + 3z = 26$$

In the *Nine Chapters* the conditions were given as columns in a matrix, reading from right to left. Then column operations were performed to simplify the matrix.

**Chinese Mathematics**

$$\begin{bmatrix} 1 & 2 & 3 \\ 2 & 3 & 2 \\ 3 & 1 & 1 \\ 26 & 34 & 39 \end{bmatrix} \longrightarrow \begin{bmatrix} 1 & 6 & 3 \\ 2 & 9 & 2 \\ 3 & 3 & 1 \\ 26 & 102 & 39 \end{bmatrix} \longrightarrow$$

$$\begin{bmatrix} 1 & 3 & 3 \\ 2 & 7 & 2 \\ 3 & 2 & 1 \\ 26 & 63 & 39 \end{bmatrix} \longrightarrow \begin{bmatrix} 1 & 0 & 3 \\ 2 & 5 & 2 \\ 3 & 1 & 1 \\ 26 & 24 & 39 \end{bmatrix} \text{ and after several } \longrightarrow \begin{bmatrix} 0 & 0 & 3 \\ 0 & 5 & 2 \\ 36 & 1 & 1 \\ 99 & 24 & 39 \end{bmatrix}$$

We can reformulate the final matrix as a set of easily solved equations

$$3x + 2y + z = 39$$
$$5y + z = 24$$
$$36z = 99$$

## CHINESE NUMERALS

Chinese numerals follow the pattern of using symbols for the numbers 1, 2, 3, 4, 5, 6, 7, 8, 9, 10, 100, 1000, .... If we let $10 = a$, $100 = b$, and $1000 = c$, we can illustrate that 532 would be written in the following manner:

$$\begin{matrix} 5 \\ b \\ 3 \\ a \\ 2 \end{matrix}$$

The actual Chinese symbols are

1 一
2 二
3 三
4 四
5 五

6 六
7 七
8 八
9 九

As was mentioned, the abacus was not used until about A.D. 1200 in China. What the Chinese used before that was a set of bamboo rods which were kept in a bag. Calculations were performed with these rods, very quickly, on a table or on the ground. Here again, a positional system was used. The following symbols represent the numerals 1–9.

| 𝟏 𝟐 𝟑 𝟒 𝟓 𝟔 𝟕 𝟖 𝟗 |

or

These two sets of symbols are used in alternate positions left to right. For example,

532

4796

The Chinese used black rods for negative values and red for positive, so they understood the idea of a negative number as representing a loss.

## CHINESE CONTRIBUTIONS TO MATHEMATICS

Chinese civilization was very advanced. Many inventions were used in China long before they were known in the West, such as printing and gunpowder (eighth century) and paper and the compass (eleventh century), but the Chinese never went on to develop science and technology in the manner of the West. The high point in Chinese mathematics (until modern times) occurred in the thirteenth century. Several mathematicians are known to have written works in this period, and perhaps many other works are lost.

Chinese Mathematics

One type of problem which was known to these men and which was done in China best and earliest was the finding of numerical roots to equations. Chinese mathematicians had a very nice algorithm which was rediscovered by Horner in the 1800s, and is now called *Horner's method*. It is studied in the theory of equations. Horner's method can be used to find higher roots, such as roots of $x^7 - 2 = 0$, or approximate solutions of more complicated equations, such as $x^5 + 3x^4 - 2x + 1 = 0$.

Another topic which occurs first in Chinese mathematics (from 1100) is the Pascal triangle (named after a Frenchman of the 1600s) which gives coefficients in binomial expansions.

$$(x + y)^0 = 1$$
$$(x + y)^1 = 1x + 1y$$
$$(x + y)^2 = 1x^2 + 2xy + 1y^2$$
$$(x + y)^3 = 1x^2 + 3x^2y + 3xy^2 + y^3$$
$$(x + y)^4 = 1x^4 + 4x^3y + 6x^2y^2 + 4xy^3 + 1y^4$$

The interesting feature of this triangle is that each row can be obtained from the preceding row by addition. For example, to get the coefficients of $(x + y)^5$, first write the coefficient row of $(x + y)^4$

$$1 \quad 4 \quad 6 \quad 4 \quad 1$$

For the next row, write a one on either side for the first and last coefficients, and in the space between and under any two numbers write their sum.

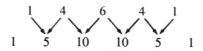

Thus, the coefficients of $(x + y)^5$ are 1, 5, 10, 10, 5, 1.

## 3 Indian Mathematics

The development of mathematics in India dating from about A.D. 300 shows definite Hellenistic influence. Books called the *Siddhantas*, or systems (of astronomy), were written, one of them being the *Surya*

*Siddhanta*, or *System of the Sun*. These works dealt with epicycles and calculations that were dependent upon trigonometry, which shows that there was probably some contact between the Indians and the Greeks of Alexandria through trade and immigration. One of the ways trigonometry was changed in the *Siddhantas* from the Greek work was to use the half-chord, or sine, function instead of the chord.

To make it easier to memorize, a sine table was given in the *Surya Siddhanta* in verse form, but since numbers have few rhymes, words were adopted for numbers. For example, in place of one the Indians wrote *moon*, because there is only one moon. For two they wrote *eyes*, *arms*, or *wings*. Thus, 5021 could have been represented in the following manner.

moon–wings–hole–senses

Note that the number was expressed with the units position first. This illustrates the use of a positional system, because the first five really represents 5000. We can be certain that by A.D. 500 the positional system was in use in India.

### INDIAN NUMERALS

There were several sets of numerals used in India, the most interesting being the Brahmi which dated from around 300 B.C. The Brahmi is similar to the alphabetical Ionic Greek system, but it is not clear what the original derivation of the Brahmi symbols was or how much, if any, Greek influence there was on the development of the symbols. The Brahmi numerals are:

| 1 | 2 | 3 | 4 | 5 | 6 | 7 | 8 | 9 |
|---|---|---|---|---|---|---|---|---|
| 10 | 20 | 30 | 40 | 50 | 60 | 70 | 80 | 90 |

Some symbols for the hundreds, but not all, have been found recorded, but we will give none here.

For some reason, by about 600 the Indians used the first nine symbols in a positional manner, as we do now, disregarding the other symbols. Perhaps the idea came from the abacus where only nine symbols are needed, or perhaps from Babylon or China where positional systems were already in use. There was contact among China, India, and the

West, but the amount of transfer of knowledge and in which direction is not really known. The zero, which is definitely needed with a positional system, was introduced a bit later, certainly by 876, as a small dot ·. In India it was called *sunya*, or *void*, which became *as-sifr* in Arabic and *zephirum* in Latin (other similar Latin forms were used). These terms led to our words *cipher* and *zero*.

## BRAHMAGUPTA (628) AND BHASKARA (1150)

Indian mathematics was very much in the Babylonian spirit with collections of various types of problems, some exact methods, and some approximate. Greek theoretical geometry had reached its limits, but India carried forward the Babylonian numerical approaches and ultimately greatly influenced the development of arithmetic and algebra in Europe in the Middle Ages. Later still, the calculus and other powerful new mathematics were created by a merging of the numerical algebra with the theoretical geometry.

Two important Indian mathematicians who wrote works on numbers, algebra, and other topics are Brahmagupta and Bhaskara. To give a specific example from their work, consider numerical rules for the solution of quadratic equations. In Brahmagupta's book the equation

$$x^2 - 10x = -9$$

is solved. It was written there as

$$ya \quad v \quad 1 \quad ya \quad 1\dot{0}$$
$$ru \quad \quad \quad \dot{9}$$

Brahmagupta used the abbreviation *ya* for the unknown, *ya v* for the square of the unknown, and *ru* for the constant. The left-hand side of the equation was written on the top line and the right-hand side on the bottom line. A negative number was indicated by a dot placed above.

The solution to the problem is given step-by-step following the plan

$$x = \frac{\sqrt{-9 \cdot 1 + (-5)^2} - (-5)}{1} = 9$$

The solution reads,

> Here ... number ($\dot{9}$) multiplied by (1) the coefficient of the square gives $\dot{9}$ and added to the square of half the coefficient of the middle term, namely 25, makes 16; of which the square

root 4, less half the coefficient of the unknown 5 is 9; and divided by the coefficient of the square, 1, yields the value of the unknown, 9.[1]

Indian mathematicians computed with their nine or ten numerals on a dust board or on paper made of palm leaves. They developed several methods of computation which were later used by Arabic and European mathematicians. It is not clear when they were first developed, or by whom, but many are found in a book by Bhaskara. His book was called *Lilavati* after his daughter, and there is an interesting story that explains why Bhaskara gave her name to the book.

Astrologers had predicted that Lilavati would never wed. Bhaskara, however, divined a lucky moment for her marriage. The ceremony was to take place at the end of an hour which was marked by an hour cup floating on a vessel of water. This cup had a small hole in the bottom of just the proper size so that the water would trickle in and sink it at the end of the hour. Lilavati, with natural curiosity, looked to see the water rising in the cup, and a pearl dropped from her clothes accidentally stopping the flow. The hour passed without the sinking of the cup, and Lilavati was thus fated never to marry. To console her Bhaskara wrote a book in her honor, promising her immortality in this way.

In his book Bhaskara gave the complementary plan of subtraction. In this method the lower digit is always subtracted from 10, and the result is added to the upper digit. The advantage of this procedure is that only the differences from 10 need to be learned. Consider

$$\begin{array}{r} 6273 \\ -1528 \\ \hline \end{array}$$

Doing the complementary subtraction step-by-step, we say

8 from 10 is 2, 2 and 3 is 5
2 from 6 is 4
5 from 10 is 5, 5 and 2 is 7
1 from 5 is 4

so the answer to the subtraction is 4745.

Bhaskara gives five methods of multiplication in *Lilavati*. The most important was later named the *grating method* (*gelosia* in Italian) because it resembled the grating put across windows of private residences, so passers-by could not look in. In this method one avoids any "carries" until the final addition. It was used until printing was developed and then was gradually phased out in the 1600s, because the grating was hard to

---

[1] Henry Thomas Colebrooke, Esq., trans., *Algebra of Brahmegupta and Bhascara*, p. 347.

print. As an example of the grating method, consider 738 × 416. To find the solution, make a grating as shown in figure 3.5a. Enter 738 along the top and 416 along the right side. Multiply 4 times 8 and enter the 32, as shown. Continue with 4 × 3 = 12, and so on. When all the entries have been made, the numbers along the diagonal are added starting at the lower right. Carrying is done where needed. The product is 307,008 (fig. 3.5b).

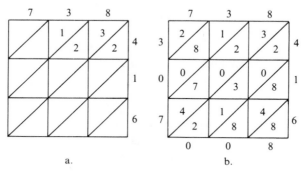

**Figure 3.5**

A Hindu method of division, very handy when work is done on a dust table but slightly confusing on paper, was later called the *galley method* because of the shape of the final result. In books, problems were decorated to look like galleys, as illustrated below. Again, the galley method lost favor with the advent of printing, but it was the most common method taught until 1600. Consider 55614 ÷ 21 as an example.

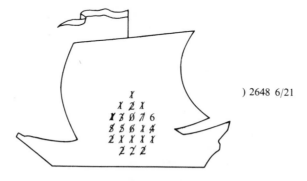

The first steps are as follows:

    55614   )2            The "guess quotient" of 55 ÷ 21 is
    21                     2.

|  |  |  |
|---|---|---|
| 1 |  |  |
| ~~5~~5614 | )2 | Starting from left to right, multiply 2 times 2, subtract that product from 5, and write the difference above, crossing out the numbers used. |
| 2~~1~~ |  |  |

|  |  |  |
|---|---|---|
| 13 |  |  |
| ~~55~~614 | )2 | Multiply 2 times 1, subtract that product from 5, and write the difference above, crossing out the numbers used. |
| 2~~1~~ |  |  |

|  |  |  |
|---|---|---|
| 13 |  |  |
| ~~55~~614 | )26 | Now move the divisor over one place and estimate the quotient of 136 ÷ 21 as 6. |
| 2~~1~~1 |  |  |
| 2 |  |  |

|  |  |  |
|---|---|---|
| 1 |  |  |
| 1~~3~~ |  |  |
| ~~55~~614 | )26 | Continuing as before, multiply 6 by 2 and subtract from 13. |
| 2~~1~~1 |  |  |
| ~~2~~ |  |  |

|  |  |  |
|---|---|---|
| 1 |  |  |
| 1~~3~~0 |  |  |
| ~~55~~~~6~~14 | )26 | Multiply 6 by 1 and subtract from 6. |
| 2~~1~~~~1~~ |  |  |
| ~~2~~ |  |  |

|  |  |  |
|---|---|---|
| 1 |  |  |
| 1~~3~~0 |  |  |
| ~~55~~614 | )264 | Move the divisor over one place and estimate the quotient of 101 ÷ 21 as 4. Continue in the same manner until a solution is obtained. |
| 2~~1~~~~1~~1 |  |  |
| ~~2~~2 |  |  |

If this problem were done on the blackboard, then the numbers could be erased rather than scratched out, and the jumbled appearance could be avoided. It would look like the following:

Indian Mathematics 109

| | | |
|---|---|---|
| 55614 | )2 | |
| 21 | | |
| 15614 | )2 | The 5 is erased and the difference, 1, written in its place. The 2 of the divisor is erased after it is used. |
| 1 | | |
| 13614 | )26 | |
| 21 | | |
| 1614 | )26 | |
| 1 | | |
| 1014 | )264 | |
| 21 | | |
| 214 | )264 | |
| 1 | | |
| 174 | )2648 | |
| 21 | | |
| 14 | )2648 | |
| 1 | | |
| 6 | | |

Of course, when the intermediate steps are erased, the problem becomes impossible to check, but some simple checks were developed as aids, and the problem can always be done over again.

As an aside, it is convenient to mention another culture in which a place-value system and zero were used. The Mayas of Yucatan, Mexico fundamentally used a base 20 system. Their symbols were the following:

$$· = 1$$
$$— = 5$$
$$⊙ = 0$$

Thus

$$17 = \begin{matrix} ·· \\ = \\ = \end{matrix}$$

110   Mathematics in Asia

Just as we write 352 for 3(100) + 5(10) + 2 the Mayas wrote (from top to bottom)

$$\overset{..}{\underset{\equiv}{\phantom{-}}}$$

$$\overset{..}{\underset{\equiv}{\phantom{-}}}$$

for 17(20) + 12. It has recently been shown that the Olmecs, another group of Indians in Mexico, originally devised this system that was later used by the Mayas. The Olmecs appeared about 1200 B.C., while 31 B.C. is the oldest date that has been attached to writings containing these numerals.

## 4  Arabic (Islamic) Mathematics

Mohammed lived in Arabia in the early 600s. After his death Mohammedanism spread to Asia, Africa, and southern Europe. Bagdad, built in about 760 by the caliph al-Mansûr, became the intellectual center of the Mohammedan world. The rulers of Bagdad encouraged learning, and many Greek works were translated into Arabic; indeed, in some cases only the Arabic version has survived. Astronomy was considered to be the most important subject of study, and the Arabic writers particularly liked the *Almagest* and the *Elements*. Because of its usefulness in astronomy, trigonometry was of great interest to the Arabs. Significant for the development of mathematics were the Arabic works on numeration and algebra which introduced these subjects to Europe.

### AL-KHOWARIZMI (825)

A most influential person was al-Khowarizmi who wrote the first work to which the name algebra was applied. Our word *algebra* comes from the Arabic *al-jabr* which seems to mean transposing. Al-Khowarizmi also wrote on arithmetic. When his works were translated into Latin, his name became *algoritmi*, from which our word *algorithm* is derived.

Al-Khowarizmi's algebra contained solutions to linear and quadratic equations. He was influenced extensively by the Babylonian and Hindu traditions, and he systematized the Babylonian treatment of quadratics reducing all such problems to the following basic types.

1. $x^2 = ax$      2. $x^2 = a$      3. $ax = b$
4. $x^2 + ax = b$  5. $x^2 = ax + b$  6. $x^2 + b = ax$

Al-Khowarizmi solved a quadratic equation by a step-by-step rule, as did the Babylonians and Indians, but then he gave a geometric diagram as a proof.

Although al-Khowarizmi was not greatly influenced by Greek mathematics, he may have derived from Euclid's *Elements* the idea that geometric proof is desirable. However, al-Khowarizmi's use of geometry was much different than that of the Greeks. He did not construct a "line" solution to his problem as was done, for example, in Book II of Euclid where the solution is *synthesized* from the given. Here is a favorite example from the Arabic work,

$$x^2 + 10x = 39$$

Al-Khowarizmi solved this equation using a rule, then he gave a diagram, as illustrated in figure 3.6, and the following explanation.

*Method*  Let a square of side $x$ be given (fig. 3.6a). On each side construct a rectangle of side $10/4 = 2\frac{1}{2}$ (fig. 3.6b).

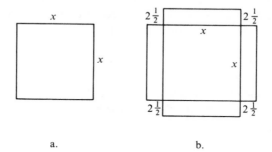

a.                    b.

**Figure 3.6**

This figure now represents

$$x^2 + 4(2\tfrac{1}{2})x = x^2 + 10x$$

the left-hand side of the equation. Now complete the square (fig. 3.7). Add to $x^2 + 10x$ the four squares of

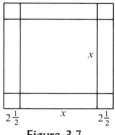

**Figure 3.7**

area $(5/2)^2 = 25/4$ each, for a total area of the large square of

$$39 + 25 = 64$$

Thus, the side of the large square is 8. Since $x + 2\frac{1}{2} + 2\frac{1}{2} = 8$, then $x = 3$.

Here al-Khowarizmi did not construct $x$, but rather started with $x$. His figure was not drawn to scale; he was *analyzing* the problem, and his approach was basically algebraic, even in his use of the diagram. He carried on the Babylonian-Indian algebra, but with the Greek influence that proof, hopefully geometric, is needed.

It is interesting that our word *root* is derived from an Arabic and Indian concept. Somehow the Arabic writers thought of a square root as analogous to a root of a plant. Thus, they used the same name for both square root and plant root. The Latin translation of root is *radix*, which is similar to *radish* and *radical*. You might say these words all have the same root.

Al-Khowarizmi also used a method of *double false position* for solution of a simple linear equation, $ax + b = 0$. The method may have been derived from Indian mathematics, and was widely used later in Europe.

*Method*  Let $g_1$ and $g_2$ be two guesses as to values of $x$ and let $f_1$ and $f_2$ be the failures, that is, the values of $ag_1 + b$ and $ag_2 + b$ (which would be equal to zero if the guesses were right). The solution is given as

$$x = \frac{f_1 g_2 - f_2 g_1}{f_1 - f_2}$$

It was often considered amazing that a correct answer could be obtained from two guesses, and it certainly is a hard way to do what is now with symbolism an easy problem. As an example, consider

$$5x - 10 = 0$$

Guess $g_1 = 3$ and $g_2 = 1$. Then, by substitution, $f_1 = 5$ and $f_2 = -5$. Therefore,

$$x = \frac{5 \cdot 1 - (-5) \cdot 3}{5 - (-5)} = \frac{20}{10} = 2$$

### ARABIC NUMERALS

In addition to translating Greek, the Arabs made translations of Indian works, such as the *Siddhantas*. Arabic writers were influenced by India with respect to numerals and trigonometry. Two Arabic systems of numerals were derived from the original Brahmi numerals, an East Arabic form used by Arabs in the Middle East

/ ⲣ ⲣⲡ ⲣⲡ ⲱ ϥ ⲩ ⲁ ⳇ

which is very similar to their present forms, and a West Arabic form used by Arabs in Spain

1 2 3 4 5 6 7 8 9

from which our numerals developed. It is not known whether the Arabs brought the numerals with them from the Middle East to Spain, or whether they found them already in use when they reached Spain. The numerals may have been transmitted to Spain via Alexandria, since it is possible that the Hindu numerals were known in Alexandria at an early date.

A theory by a Russian historian, Bubnov, asserts that the West Arabic numerals were derived from numerals used by Latin abacists to mark the counters used on the abacus.[2] Further, this theory gives an original meaning for the numerals. Unfortunately, there is not enough information available to completely trace the origin of our numerals with any claim to certainty.

I repeat this theory of the origin of the abacus numerals, because, even if not verifiable, it is an interesting explanation of the origin of the symbols. The Latin abacus symbols were

The 2 is a cursive way of writing $=$. Similarly, 3 is $\equiv$ written without removing the pen from the paper. The 4, ᖷ , is said to come from

---

[2] See Harriet P. Lattin, "The Origin of our Present System of Notation According to the Theories of Nicholas Bubnov," *Isis*, pp. 181–94.

writing $-\mathrm{I}-$ in one motion. The 5, $\varepsilon$ , comes from $|\Xi$ . The numerals for 6 to 9 are diacritical signs. The 6, $\triangleright$ , indicates one more than 5. The 7, $\triangledown$ , indicates two more than 5. The 8 is a closed $\mathfrak{Z}$ , and the $9$ , being four more than 5, resembles a 4.

In any case, Arabic writers used the previously mentioned West Arabic and East Arabic forms of numerals and gave methods of computation learned from India. The Arabs were familiar with both the Greek chord and the Indian sine approach to trigonometry, but they adopted the Indian approach. Europe learned Indian trigonometry from the Arabic writer **al-Battani (900)**. **Abu'l-Wefa (970)** gave a sine table, and used all six common trigonometric functions — sine, cosine, tangent, cotangent, secant, and cosecant. He also did some writing on algebra.

A very interesting figure is the Persian **Omar Khayyám (1100)** who is quite famous as the poet who wrote the *Rubaiyat*. He wrote on algebra, not only solving the usual quadratics, but also giving geometric solutions of cubics via intersecting conics. He classified cubic equations according to the number of terms as *simple*

$$r = x, \quad r = x^2, \quad r = x^3, \quad ax = x^2, \quad ax = x^3, \quad ax^2 = x^3$$

and *compound*

a. $x^2 + bx = c, \quad x^3 + bx^2 = cx$, etc.

b. $x^3 + d = bx^2 + cx$, etc.

This classification seems natural since these all seem to be different problems. After all, a cube equal to a number is not the same as a cube equal to a multiple of a side, is it? Symbolism with literal coefficients and the possibility of the coefficients being zero was needed before classification by degree could be perceived.

Several Arabic mathematicians made important contributions to mathematics. One area that has not been mentioned is geometry. For example, Omar Khayyám tried to prove Euclid's fifth postulate about parallels. Others had tried this, of course, but he made a particularly good attempt. There were noteworthy Arabic mathematicians who lived after Omar Khayyám, but by 1450 Europe had become the leading center of mathematics.

## Problems

1. Using an abacus, perform the following operations. (If you do not have an abacus available, you might use checkers, buttons, or cardboard markers.)
   a. 47 + 68
   b. 695 + 527
   c. 3 × 17
   d. 49 × 84
   e. 37 × 105
   f. 568 × 762

2. Construct a 4 × 4 magic square by writing the numbers in order starting with the number 1 in the upper left corner. Then replace each number not on one of the two diagonals by its difference from 17. Check to see that each row, column, and diagonal has the same sum.

3. Complete all steps of the solution of the 3 × 3 system of equations from the *Nine Chapters on the Mathematical Art*.

4. Write the following numbers using the Chinese system, but using our numerals, with the additional symbols $a = 10$, $b = 100$, and $c = 1000$.
   a. 35
   b. 249
   c. 307
   d. 80
   e. 6342
   f. 5004

5. Expand $(x + y)^8$ without multiplying.

6. The Chinese rods were often used on ruled tables on which a zero could be clearly indicated by leaving a blank column. Sometimes numbers were written in the form of the rod numerals, so a written zero, 0, was developed. Express the following numbers using the rod numerals.
   a. 35
   b. 249
   c. 307
   d. 80
   e. 2897
   f. 4069

7. Write the following equations using the symbolism of Brahmagupta.
   a. $3x^2 - 5x = 7$
   b. $2x^2 = 4x - 3$

8. State Brahmagupta's method of solution for the equation $2x^2 + 3x = 2$.

9. Solve the following using the complementary plan of subtraction. Explain each step.
   a. $\phantom{-}682$
      $-291$
   b. $\phantom{-}4532$
      $-2784$

10. Multiply the following numbers using the grating method.
    a. 32 × 64
    b. 675 × 238
    c. 942 × 926
    d. 68 × 4317

11. Perform the following divisions using the galley method of division.
    a. $638 \div 27$   b. $7942 \div 38$   c. $49{,}172 \div 321$   d. $671{,}411 \div 19$

12. Write the following numbers using the Mayan symbols.
    a. 25      b. 194      c. 301      d. 58

13. Give a geometric solution in the manner of al-Khowarizmi to the following equations.
    a. $x^2 + 6x = 16$      b. $x^2 + 24x = 25$

14. Use the rule of double false position to solve the equation
    $$3x - 12 = 0.$$

15. Prove that the value $x = (f_1 g_2 - f_2 g_1)/(f_1 - f_2)$ obtained using the method of double false position to solve the equation $ax + b = 0$ is correct.

16. The method of double false position can be used to find approximate solutions to equations of degree higher than one, such as $x^2 - 3 = 0$. For the equation $x^2 - 3 = 0$ the solutions approach the exact solution, $\sqrt{3}$.
    a. Let $g_1 = 1$ and $g_2 = 2$ be the two guesses. Show that the failures are $f_1 = -2$ and $f_2 = 1$, and find $x$.
    b. Repeat the rule of false position. This time use the $x$ of part a as one guess. As the second guess use one of the guesses in part a, namely, the one whose failure has a sign opposite that of the failure of $x$. The new solution will be yet a better approximation to $\sqrt{3}$.

Some problems in the Indian books are quite fanciful. Problems 17–19 are from the *Vija-Ganita* of Bhaskara.

17. "[Arjuna,] the son of Pritha, exasperated in combat, shot a quiver of arrows to slay Carna. With half his arrows he parried those of his antagonist; with four times the square root of the quiver-full, he killed his horse; with six arrows he slew Salya; with three he demolished the umbrella, standard and bow; and with one he cut off the head of the foe. How many were the arrows, which Arjuna let fly?"[3] [Let $x^2$ represent the total number of arrows.]

18. "The eighth part of a troop of monkeys, squared, was skipping in a grove and delighted with their sport. Twelve remaining [monkeys] were seen on the hill, amused with chattering to each other. How many were they in all?"[4]

---

[3]Colebrooke, *Algebra*, p. 212.
[4]Ibid., pp. 215–16.

**19.** "The square root of half the number of a swarm of bees is gone to a shrub of jasmin; and so are eight-ninths of the whole swarm; a female is buzzing to one remaining male that is humming within a lotus, in which he is confined, having been allured to it by its fragrance at night. Say, lovely woman, the number of bees."[5] [Let $2x^2$ be the number of bees.]

**References**

The Abacus

    Menninger                      Smith (1)

Chinese Mathematics

    Midonick                       Struik (1)
    Smith (1)

Mathematics in India

    Colebrooke                 Neugebauer (1)
    Karpinski                      Smith (1)
    Midonick
    National Council of Teachers of Mathematics

Arabic (Islamic) Mathematics

    Boyer (1)                      National Council of Teachers
    Gandz (1, 2)                     of Mathematics
    Karpinski                      Smith (1)
    Lattin                          Struik (2)
    Midonick

---

[5]Ibid., pp. 211–12.

# 4 European Mathematics Until 1630

In the early medieval period there were few mathematical works available in Europe. By the 1100s more books were appearing, primarily in translations from the Arabic. Europeans learned arithmetic and algebra from these Arabic sources and developed those subjects further. At the same time, Greek works were slowly increasing in availability and influence. At first they were translated from Arabic copies, but in the Renaissance more complete Greek versions were recovered. Finally, by about 1600, mathematicians were able to master the Greek works, and with the newly developed symbolic algebra, they continued the advancement of mathematics to an even higher level.

## 1 Europe in the Middle Ages (529–1436)

Rome had been a mighty power for hundreds of years, but by A.D. 500 the central government found itself unable to maintain control of such a large empire. Communities became isolated to a degree, because the society was rural and agricultural, eventually evolving into feudalism. Vassals tilled the land of their feudal lord, working as tenant farmers. They pledged their fealty to the lord and served as his knights in return for their small parcels of land. Feudal societies were not conducive to formal education, so the only important centers of learning in the early Middle Ages in western Europe were the monasteries.

## BEDE (700)

One of the most important early mathematical figures, Bede, was an English monk. He was concerned with computing the correct date for the festival of Easter, one of the problems which faced the monks in the monasteries. Calculation was also an important practical need.

Bede recorded information about finger reckoning, a system of calculation by sign language which was widely used before and after Bede's time. Finger reckoning was a valuable skill, because people could communicate purely by means of hand signs. There was no need to write or to be able to speak the same language as long as the finger reckoning symbols were understood. Finger reckoning could also be used to record figures temporarily while calculating on an abacus or dust table. The commonly used symbols are the following:

| Number | Representation |
|--------|----------------|
| 1 | fold little finger of left hand |
| 2 | fold ring and little fingers |
| 3 | fold middle, ring, and little fingers |
| 4 | fold middle and ring fingers |
| 5 | fold middle finger |
| 6 | fold ring finger |
| 7 | fold little finger across palm |
| 8 | fold ring and little fingers across palm |
| 9 | fold middle, ring, and little fingers across palm |

The numbers 10, 20, 30, 40, 50, 60, 70, 80, and 90 are shown by using the thumb and index finger of the left hand. Thus, all numbers from 1 through 99 can be represented on the left hand. Larger numbers can be represented on the right hand. Juvenal, a Roman poet, refers to this when he says, "Happy is he indeed who has postponed the hour of his death so long and finally numbers his years upon his right hand."[1]

There was little mathematical activity from the time of Bede to that of **Gerbert (980)**. Gerbert, who later became Pope Sylvester II, wrote on computation using the abacus and on Hindu-Arabic numerals, which the may have learned during a trip to Spain. He made counters with these numerals inscribed on them and gave a method of division using

---

[1]D. E. Smith, *History of Mathematics*, II, p. 197.

such counters on an abacus. The following example is a written form of his method.

**Example** Consider $805 \div 7$. Write 7 as $10 - 3$. Dividing 10 into 805 gives 80. Multiplying, $80 \times 10 = 800$ and $80 \times 3 = 240$. The division continues as illustrated.

$$80 + 20 + 10 + 3 + 1 + \frac{7}{7} = 115$$

$$
\begin{array}{r}
10 - 3\overline{)805} \\
800 \\
\underline{\phantom{00}5} \\
+240 \\
\underline{245} \\
200 \\
\underline{\phantom{0}45} \\
+\phantom{0}60
\end{array}
\quad
\begin{array}{r}
105 \\
100 \\
\underline{\phantom{00}5} \\
+\phantom{0}30 \\
\underline{\phantom{0}35} \\
30 \\
\underline{\phantom{00}5} \\
+\phantom{00}9
\end{array}
\quad
\begin{array}{r}
14 \\
10 \\
\underline{\phantom{0}4} \\
+\phantom{0}3 \\
\underline{\phantom{0}7}
\end{array}
$$

## THE TWELFTH CENTURY

The 1100s brought a new interest in learning to Europe. Trade was increasing and towns were developing around the trading centers. The economy was tending to be based on money, so the feudal landowners were no longer the most powerful men. The power belonged to the wealthy men who could pay their armies with money, rather than land. Schools in the cathedrals were growing, and in the latter part of the twelfth century several such cathedral schools developed into some of the first universities — Paris and Oxford, for example. *University* originally meant *group* as in a *guild*, because the students were united to prevent their exploitation by local landlords and shopkeepers who could charge outrageous prices, due to the large demand for goods and services from the students in the town, unless checked.

Few books were available at this time in Latin, although there were poor versions of Boethius' mathematics. The Moslems, who were the most learned men of the times, had been in Spain and Sicily for 300 or 400 years, so scholars traveled to Spain, which was very cosmopolitan, to further their education. There were Christians, Moslems, and Jews, all of whom were important in Spanish cities, particularly Moslems and Jews when the area was under Moslem control. Many Arabic and some Greek works existed in Spain, and scholars translated these works to Latin. For example, Euclid's *Elements* was translated by **Adelard of Bath** in 1142, Ptolemy's *Almagest* was translated by **Gerard of Cremona** in 1175, and al-Khowarizmi's *Algebra* was translated by **Robert of**

Chester in 1145. In some cases an Arabic work was itself a translation from Greek, so it is understandable that much was lost in the subsequent translation.

The algebra of al-Khowarizmi was particularly liked by Europeans. It introduced them to numerical algebraic solutions to quadratics and to computation with Hindu-Arabic numerals. We shall later see some of these early European methods of computation with numerals, many of which were learned from al-Khowarizmi and other Arab writers.

## FIBONACCI (1220)

The Italian cities, particularly Venice, were the leaders in trade, and the son of an Italian trader became known as the best mathematician of the Middle Ages. Leonardo de Pisa, better known as Fibonacci (son of Bonaccio), traveled with his father to the Middle East and Africa learning various methods of computation with Hindu-Arabic numerals and possibly some Chinese mathematics. These methods were quite useful in trade, but officials familiar with the counter abacus and Roman numerals were skeptical and banned the use of Arabic numerals in Venice. One argument that the officials used was that Arabic numerals were easily altered.

Fibonacci wrote a book called *Liber abaci* (book of the abacus) containing numerals and computation. It might be called the book of calculation. The following is an example of the borrowing and repaying plan of subtraction which Fibonacci probably learned from the Arabs. The number borrowed is added to the bottom digit in the next column.

$$\begin{array}{r} 6354 \\ -2978 \\ \hline \end{array}$$

8 from 14 is 6
8 from 15 is 7
10 from 13 is 3
3 from 6 is 3

Fibonacci did not use the Hindu-Arabic numerals for fractions. He used unit fractions and common fractions for commercial problems, and he used sexagesimal fractions in theoretical problems.

Fibonacci is most famous for the *Fibonacci numbers* which come from the following problem.

> How many pairs of rabbits will be produced in a year, beginning with a single pair, if in every month each pair bears a new pair which becomes productive from the second month on?

We can find the rabbit population in each month by adding the number of pairs in the previous month to the number of new pairs born (equal to the number of pairs two months before). Thus, $u_n = u_{n-1} + u_{n-2}$, where $u_n$ is the number of pairs in month $n$. Consider the following table.

| Month | 1 | 2 | 3 | 4 | 5 | 6 | 7 | 8 | 9 | ... |
|---|---|---|---|---|---|---|---|---|---|---|
| Pairs of rabbits | 1 | 1 | 2 | 3 | 5 | 8 | 13 | 21 | 34 | ... |

Fibonacci numbers are one of the most fascinating topics in mathematics. There is at present an entire journal called the *Fibonacci Quarterly* devoted to these numbers and similar topics. There are some interesting facts about Fibonacci numbers which were discovered later in history. For example, a simple theorem about Fibonacci numbers is that every fifth number is divisible by 5. As we see from the table, $u_5 = 5$ and $u_{10} = 55$. Another interesting property concerns the ratios of successive Fibonacci numbers,

$$\frac{1}{1}, \frac{1}{2}, \frac{2}{3}, \frac{3}{5}, \frac{5}{8}, \ldots$$

This sequence of ratios converges to the golden section number $(\sqrt{5} - 1)/2$. Recall that the golden section number $x$, satisfies the relation

$$\frac{1}{x} = \frac{x}{1 - x}$$

$$1 - x = x^2 \text{ (fig. 4.1)}$$

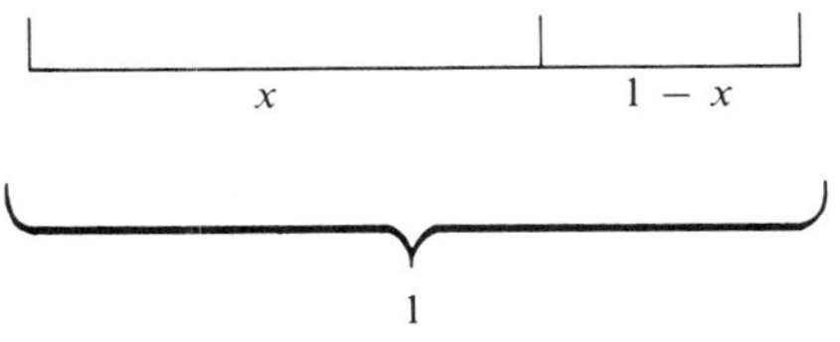

**Figure 4.1**

To illustrate the relation between this value $x$ and the Fibonacci numbers, continued fractions are used. An example of a continued fraction is

$$\cfrac{3}{2 + \cfrac{1}{4 + \cfrac{1}{3}}}$$

which can be simplified to give

$$\cfrac{3}{2 + \cfrac{1}{\frac{13}{3}}} = \cfrac{3}{2 + \cfrac{3}{13}} = \cfrac{3}{\frac{29}{13}} = \frac{39}{29}$$

Consider the simplest infinite continued fraction

$$x = \cfrac{1}{1 + \cfrac{1}{1 + \cfrac{1}{1 + \cfrac{1}{1 + \cdots}}}}$$

If you examine the fraction carefully, you can see that

$$x = \frac{1}{1 + x}$$

so that

$$x + x^2 = 1$$

and $x$ is, in fact, equal to the golden section number. Take successive approximations to the infinite continued fraction by chopping it off.

$$x_1 = \frac{1}{1}$$

$$x_2 = \cfrac{1}{1 + \frac{1}{1}} = \frac{1}{2}$$

$$x_3 = \cfrac{1}{1 + \cfrac{1}{1 + \frac{1}{1}}} = \cfrac{1}{1 + \frac{1}{2}} = \cfrac{1}{\frac{3}{2}} = \frac{2}{3}$$

and so on. Thus, the approximations to the continued fraction are just the ratios of successive Fibonacci numbers. Since this continued fraction equals the golden section number, the sequence $1/1, 1/2, 2/3, 3/5, \ldots$ converges to the golden section number.

The sequence of Fibonacci numbers occurs often in nature. For example, the whirls of the rings on a pineapple are given by Fibonacci numbers. The hexagonal regions are arranged in rows in various directions — five parallel rows sloping gently up to the right, eight rows sloping more steeply to the left, and 13 rows sloping very steeply to the right. (All rows are started at the base of the pineapple.) Notice that the numbers of rows — 5, 8, and 13 — are from the Fibonacci sequence. A similar phenomenon occurs in the patterns of sunflower seeds and fir cones.

## METHODS OF COMPUTATION

**Maximus Planudes (1300),** a Greek monk from Constantinople, used a division method which is similar to our own, but still related to the galley method in that the divisor is multiplied digit-by-digit. Consider the problem 625 ÷ 25. Divide 25 into 62, giving 2. Multiplying, 2 × 2 = 4. Subtract the 4 from the 6 and bring down the 2, giving 22. Now, 2 times 5 is 10, and subtracting gives 12. Bring down the five, and divide 25 into 125, giving 5. Multiply and subtract as before.

```
       25)  625   (25
             4
            ___
            22
            10
            ___
            125
             10
            ___
             25
             25
            ___
```

John of Halifax (1250) was known as **Sacrobosco,** the Latin translation of Halifax, because he wrote in Latin. Sacrobosco was educated at Oxford, and he taught in Paris. The simple borrowing plan of subtraction that is familiar to us was found in a popular book on computation written by Sacrobosco.

```
     42       7 from 12 is 5
    -27       2 from  3 is 1
    ___
     15
```

Early methods for the basic operations of arithmetic tended to follow patterns suitable for use on a dust board (as the Indians used), even when done on paper. For example, the following is an addition from Sacrobosco.

```
   826    829    909    1309
   483     48      4
```

On a dust table or slate the figures would not be copied over as they are here, but rather be erased. It was not until the appearance of printed arithmetic books in the fifteenth century that addition assumed its modern form.

An anonymous work, *The Crafte of Nombrynge* (1300), is the first arithmetic book to be written in English. The following computation taken from this book is an example of a multiplication. The calculation is really more suitable for writing on a dust board or slate than on paper, so to make it clearer numbers which would be erased will have slashes through them. The approach is similar to that in the galley method of division.

**Example**   Compute 2465 × 432.

    2465
    432

      4        Take the 2 of 2465 and multiply by the 4 in 432, writing
86̸2465   the 8 above and crossing out the 4. Then 2 × 3 is 6
4̸3̸2̸      (written above 3), and 2 × 2 is 4 (written above 2).

   12      Move the 432 over one place to the right, and repeat the
  1648    process, multiplying 432 from left to right by 4.
86̸2̸465
4̸3̸2̸2̸
  4̸3̸

   21      Continue in this way until each digit of 2465 has been
   10      used. To obtain the product, add the figures from left to
 2411    right, carrying back to the left when necessary. The final
 1285    result is shown at the left.
164820
86̸2̸46̸5̸
4̸3̸2̸2̸2̸2̸
 4̸3̸3̸3̸
   4̸4̸
―――――
9̸5̸3̸880
 1064

Addition was done from left to right. When figures were scratched out, as was done here, it was called the scratch method.

Since we are discussing methods of computation, it is convenient to mention an interesting method of multiplication related to the Egyptian method of doubling and halving. It is called the *Russian peasant method* and supposedly was used by Russian peasants even in the twentieth century.[2] Consider 49 × 28. We halve 49, neglecting fractions, and double 28:

| 49 | 24 | 12 | 6 | 3 | 1 |
|----|----|----|----|----|----|
| 28 | 56 | 112 | 224 | 448 | 896 |

[2]Ibid., p. 106.

We now add the figures in the bottom row under odd numbers in the top row,

$$\begin{array}{r} 28 \\ 448 \\ \underline{896} \\ 1372 \end{array}$$

We are, in effect, adding one 28 to sixteen 28s to thirty-two 28s.

## BRADWARDINE (1325) AND ORESME (1360)

In the medieval universities there were several scholars who contributed to mathematics. Two of the most important writers were **Thomas Bradwardine** of Oxford who became Archbishop of Canterbury and **Nicole Oresme** of Paris who became Bishop of Lisieux.

In their criticisms of the ideas of Aristotle on motion they extended the idea of proportions of Euclid to talk about $y \propto x^n$ or even a proportion anticipating the later development of logarithms. In the latter case one might have a sequence which varied geometrically, while a corresponding sequence varied arithmetically

| $x$ | 1 | 2 | 4 | 8 | 16 | ... |
| $y$ | 1 | 2 | 3 | 4 | 5  | ... |

In the geometric approach these men used, these ideas are very complicated to express. What makes it easier for us is a good symbolism which they did not possess.

The basic concepts of Aristotelian and medieval physics had been qualities. A body was hot or cold, heavy or light, red or black, etc. An object would behave in a certain way because of its properties; if an object was heavy, it would fall to the ground. By the late Middle Ages scholars began discussing variable qualities such as velocity or temperature. In this study they made early attempts at graphs and struggled with the equivalent of infinite series. For example, Oresme drew a triangle representing a body moving with increasing velocity at a constant rate (fig. 4.2). The horizontal direction represents time, while the vertical represents velocity. He found the distance covered as the area of the triangle.

There were various speculations on the infinite and infinitesimals, although we will not be concerned with them in this text. Several hundred years later men like Galileo, who created many concepts of quantitative physics, read these medieval works and were influenced by them.

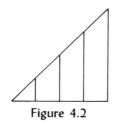

**Figure 4.2**

Oresme studied incommensurables extensively, because he wanted to show that the planets' motions were given by irrational ratios, such as the ratio of the velocities of orbit of two planets, and thus were unknowable, to combat the influence of the astrologers on his king, Charles V. Charles was a firm believer in astrology, and Oresme was possibly afraid that the king would make unwise decisions based on the astrologers' predictions.

In the middle of the fourteenth century, the terrible black death swept over Europe, and nearly half of the population died. This plague greatly interrupted the continuity of mathematical learning. England and France were also at war for the next hundred years, a situation which further quelled mathematical activity for some time.

## 2  The Renaissance

Renaissance refers to the rebirth of interest in learning by the mid 1400s, at which time Europe took the lead in the development of mathematics. There were several factors to account for Europe's importance in mathematics. Printing was introduced in Europe and had a tremendous effect on learning. Books became more plentiful and, therefore, more accessible to more people. A rag paper process was developed at the same time. Previously, one had to write on papyrus or parchment. The rag paper added to the usefulness of printing, because more books could be printed. It was not until the 1800s that a cheap wood pulp paper process was developed.

Trade grew considerably in the Renaissance. An improved rudder was put into use, enabling ships to travel even greater distances. Italy, the former leader in trade, was still very strong in the early Renaissance. German groups were important in trade in the northern European countries, and they formed the Hanseatic League, a group devoted to protecting German trading interests in foreign countries. Schools were established by the Hanseatic League and the *rechenmeister* (master of calculating) became a kind of town official. The Hanseatic schools can be thought of as business schools for future traders. The teachers wrote many commercial arithmetics for use in the schools.

Because trade was increasing, there was also a need for astronomical tables for navigation (to calculate longitude, for example) which, in turn, required trigonometric tables. Many people were interested in astrology and this influenced the demand for astronomical tables, since the precise locations of the planets at any time were needed to make astrological predictions.

The Renaissance fostered a renewed interest in classical authors. The literary scholars wanted to study original Latin and Greek works, including mathematics texts. Because of the search for the original books, many works not previously known were eventually made available, such as those of Apollonius, Diophantus, and parts of Archimedes. Yet in this period, especially the early years, these works were too advanced to be understood by many. The Arabic algebraic tradition was developed and studied more, because it was easier to understand.

## REGIOMONTANUS (1460)

Regiomontanus is the name given to Johann Müller of Königsberg, because Regiomontanus is the Latin translation of Königsberg (king's mountain). Probably the most capable mathematician of his time, Regiomontanus completed a new Latin version of Ptolemy's *Almagest*. He traveled and searched for ancient mathematics books and found Diophantus' *Arithmetic* in Greek, but never did translate or publish it.

Regiomontanus' main contribution to mathematics was a book on triangles. In it he showed the methods of finding one side of a triangle when other sides and angles are given, among many other problems. His book is amusing in some respects. For a simple problem, such as given $A = 5C$ and $B = 2C$, the method of finding $A - B$ is to subtract, $5 - 2 = 3$, and multiply by $C$. Regiomontanus goes through quite a long proof in the Euclidean style, but ultimately gives the method in a simple rule. It appears that he just wanted to use the rule, but felt he must give a proof, an approach similar to that of al-Khowarizmi. In the same text on triangles Regiomontanus expressed the lengths and sides of triangles by numbers, and if a side length turned out to be an irrational number, he merely took an approximation. Regiomontanus also developed some astronomical tables which were used by Columbus on his trip to America.

## SYMBOLISM AND NOTATION (1484-1545)

Concerning the notation for equations, Diophantus used abbreviations for unknowns and powers, as did Brahmagupta, but al-Khowarizmi wrote everything out in words, including numbers. During the Renaissance several authors wrote algebra books introducing notation and symbols, and, gradually, abbreviations and symbols evolved in Europe, so that by the mid 1600s the form of equations was as we know it now.

**Nicolas Chuquet (1484)** of France wrote *Three Parts of the Science of Numbers*. He used *plus*, meaning *more*, for addition and abbreviated it by $\bar{p}$. *Moins*, meaning *less*, was applied to subtraction and abbreviated as $\bar{m}$ (Fibonacci had used minus). Chuquet represented $6x^2$ by $.6^2$, and he wrote negative exponents such as $9x^{-2}$ as $.9^{2.\bar{m}}$. He was one of the first mathematicians to use the laws of exponents.

**Luca Pacioli (1494)** of Italy wrote a popularly used book, *Summa* . . . , which covered arithmetic, algebra, geometry, and bookkeeping. He borrowed from other works, giving several methods of multiplication and division. Among them were old methods derived from the Arabs and Hindus, but also some methods that are basically what we presently use. It was perhaps in the seventeenth century that our present methods of multiplication and division became standardized. Printing ultimately rendered useless the methods unsuitable to be set in type.

Our method of multiplication was called the *chessboard method* by Pacioli, because of its appearance. For example,

```
        9 4 3 7
          2 8
       ┌─┬─┬─┬─┬─┐
       │7│5│4│9│6│
     ┌─┼─┼─┼─┼─┘
     │1│8│8│7│4│
     └─┴─┴─┴─┴─┘
       2 6 4 2 3 6
```

An older form of multiplication from a manuscript dated 1424 is illustrated by the following example of 34 × 45.

1 5 3 0

Among the other methods given by Pacioli was that of *cross multiplication*. This system is really only useful for small problems, for example,

Cross multiplication allows the problem to be calculated mentally with only the answer being written on paper. The thought process involved is illustrated by the following steps.

$5 \times 4 = 20$; put down 0 and carry the 2.

Following the lines of the cross,

$5 \times 3 = 15$, plus $2 \times 4 = 8$, is $23 + 2$ (carried) $= 25$; put down the 5, carry 2.

Then $3 \times 2 = 6$, plus 2 (carried) is 8.

Thus, the answer is 850.

The use of the cross in this method of multiplication probably led to its later use as a symbol for multiplication.

Pacioli gave the name *a danda*, meaning *by giving*, to our division method, because at each step the next number is brought down and "given" to the remainder. This name has not survived, because we do not have to distinguish different methods of division as Pacioli did.

**Johann Widman (1490)** of Germany was the first to use the signs + and − in a text. These symbols had been in use as warehouse marks to indicate excess (+) and deficiency (−). The + sign is a ligature for *et*, meaning *and;* it is somewhat similar to an ampersand, &. The − sign may be a simplification of $\overline{m}$ which was previously used to indicate minus in other writings.

Another German, **Adam Riese (1524)**, was responsible for a table of square roots. To find $\sqrt{3}$ would be difficult without the use of decimals. An alternative notation would be

$$\sqrt{3} = \frac{\sqrt{30{,}000}}{100} \quad \text{or} \quad \frac{\sqrt{3{,}000{,}000}}{1000}$$

Riese wrote the answer as 1732 with the understanding that it was to be divided by 1000. The idea of decimals was becoming clearer, but decimals were not yet in use as we know them.

**Christoph Rudolff (1525)** was also a German mathematician. He contributed yet another step in the advancement of the concept of decimals; for example, he wrote $50\frac{3}{10}$ or 50|3 to symbolize 503 divided by 10. A Persian writer, al-Kashi (1400), also used decimals somewhat similarly.

Rudolff introduced the symbol $\sqrt{\phantom{x}}$ for square root. It was a lowercase *r*, an abbreviation for *radix*. Sometimes a capital R was used in front of the number, as R3, to indicate a square root, but by using a small letter the number can be written underneath, as $\sqrt{\phantom{x}}$ 3, which later became $\sqrt{3}$.

In the German books, algebra was known as the cossic art. *Coss* was a translation of the Latin *res*, meaning *thing*. We now call the *thing* the *unknown*.

The symbol = for equivalence was introduced by an Englishman, **Robert Recorde (1545),** who wrote several mathematical books. He used parallel lines ======= as a symbol for equality, because "no two things could be more equal."

The notations of algebra were becoming standardized in the late 1500s. An Englishman, **Thomas Harriot (1600),** wrote an algebra book in which he introduced the sign < for less than and the sign > for greater than. Harriot later went as a surveyor to the colony of Virginia.

Related to the symbolism and notation being formulated were the terms taken from Latin arithmetic which we currently use. The following table lists some of the terms commonly used in mathematics.

| Term | Meaning |
| --- | --- |
| minuend | number to be diminished (*numerus minuendus*) |
| subtrahend | number to be subtracted |
| numerator | numberer |
| denominator | namer |
| quotient | how many? |
| fraction | to break (*frangere*)<br>Fractions were called broken numbers or fractured numbers in English. |
| degree | steps (*de* + *gradus*) |
| minute | first small part (*pars minuta prima*) |
| second | second small part (*pars minuta secunda*)<br>This refers to the division of the unit into sexagesimal parts. |
| per cent | per hundred (*per cento*)<br>This was abbreviated as per c̊, 0/0, or %. The term was originally used to mean just *per hundred*, so one could say $5 per cent.[3] |

---

[3]Ibid., pp. 249–50.

The derivation of the word *tally* is an interesting one. Tally comes from the same root as *tailor* which means *to cut*. Originally, a tally was a piece of wood on which notches were cut to represent numbers. The tally was used to keep accounts by being split in such a way that the notches would appear on both halves, therefore producing two copies, or records, of a transaction. For example, when a person deposited money in the Bank of England the amount of the deposit was recorded on a tally, the tally was split, and the bank kept the *leaf* and the depositor kept the *stock*. The depositor thus became a *stockholder* in the bank![4]

## MATHEMATICAL INFLUENCE ON ASTRONOMY AND ART

**Nicholas Copernicus (1530)** lived in a copper-mining town in Poland. Copernicus is the Latin form of his Polish name, Koppernick. In his travels, Copernicus may have met some Neoplatonists in Italy who believed that mathematics was fundamental to an understanding of the universe. Copernicus was interested in astronomy, and he applied his mathematical ideas to his study of the solar system and stars. He believed that circles were the pure, eternal forms and that uniform motion was proper for heavenly bodies.

Copernicus' goal was to improve the Ptolemaic system by considering the bodies to have uniform motion with respect to the center of the epicycle rather than the equant. To achieve this goal, he found it desirable to have the earth move around the sun. Some ancient Greeks had already proposed this hypothesis, but it had not been accepted. Copernicus actually made the epicycle theory more complicated, but in approximately 50 years his theory gained support. Experimentally, there was no reason to change the hypothesis; in fact, there were arguments, such as the fact that no parallax had ever been observed, that suggested that the earth did not move.

Mathematics had an influence on art in the Renaissance. Previously, paintings were flat and more symbolic than real in appearance. Then, some mathematically inclined artists such as **Leon Battista Alberti (1435)** and **Piero della Francesca (1478)** in Italy, and **Albrecht Dürer (1525)** in Germany, began investigating the geometry of perspective. Their goal was to represent depth on the canvas.

The act of seeing is very involved, part of it being psychological; we see what we know we should see. An artist represents a scene on a flat canvas in such a manner that we perceive depth. Light travels in rays, so we can get an idea of a correct representation by seeing where the rays from our eye to an object intersect the canvas. For example, consider the

---

[4]Ibid., p. 194.

representation of a flat table top (fig. 4.3). The line $A'C'$ is the exact representation on the canvas of the table edge $AC$ as perceived from the point shown. The figure $A'B'D'C'$, though not rectangular, does convey the correct image of a three-dimensional table. It is as if the canvas were a transparent window between the eye and the object. The rays from the eye to the back corners of the table intersect the canvas in a shorter line than do the rays to the front corners.

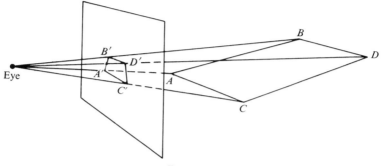

Figure 4.3

It has been said that Greek art is tactile in nature as compared to Renaissance art which is visual and that this comparison holds, in a sense, for their respective mathematics. In Greek art figures were distinct with little or no overlapping, while Greek theorems were distinct entities. In Renaissance art the visual scene was represented with much overlapping, and in Renaissance mathematics there was more of an attempt to relate different theorems in general approaches.[5]

## SOLUTIONS OF CUBIC EQUATIONS

In the early 1500s the Italian algebraists made a major advance. Methods for solving quadratic equations had been known for thousands of years, but no analogous rules had yet been developed for any higher degree equations. In 1515 **Scipione del Ferro** found an algebraic solution for one type of cubic equation. Such a method was quite valuable to a mathematician for he could challenge others to a contest in solving cubics with the expectation that he would win handily. With the prestige from such a victory, a mathematician might obtain a nice position.

**Antonio Maria Fiore** learned of the solution of the cubic from his teacher, del Ferro, and in 1535 challenged another mathematician, **Niccolo Tartaglia,** to a contest of solving cubics. At the last minute,

---

[5]See William Ivins, *Art and Geometry*, 113 pp.

Tartaglia figured out the solution to the cubic, but he said nothing. Apparently, Tartaglia wanted to keep his discovery a secret so he could use it in future encounters. However, his friend **Girolamo Cardano** persuaded him to divulge it, whereupon Cardano published the solution in 1545 in his book, *The Great Art*. Even though Cardano gave credit to Tartaglia for the solution, Tartaglia was quite upset that Cardano disclosed it. In addition to Tartaglia's solution of the cubic, Cardano solved 12 other cases of cubic equations himself. Cardano also showed how to solve the quartic, using a procedure developed by his pupil **Ferrari (1545)**.

The following is an example of the method for solving cubics given in *The Great Art* (modern symbols are used, though Cardano wrote everything out).

**Example**  Solve $x^3 + 6x = 20$ by letting $x = u - v$ and substituting in the equation, giving

$$(u - v)^3 + 6(u - v) = 20$$

Multiplying,

$$u^3 - 3u^2v + 3uv^2 - v^3 + 6u - 6v = 20$$

Choose the product $uv$ to be $1/3$ the coefficient of $x$,

$$uv = \frac{1}{3}(6) = 2$$

This choice of $uv$ makes several terms drop out. Substituting, we obtain

$$u^3 - 6u + 6v - v^3 + 6u - 6v = 20$$

or

$$u^3 - v^3 = 20$$

Now,

$$v = \frac{2}{u}$$

so

$$u^3 - \frac{8}{u^3} = 20$$

or

$$(u^3)^2 - 20u^3 - 8 = 0$$

Using the quadratic formula, $u^3$ can be found. By taking the cube root, $u$ can be found. Then substituting $u = 2/v$ in the equation $u^3 - v^3 = 20$ gives the quadratic equation $(v^3)^2 + 20v^3 - 8 = 0$ which can be solved for $v^3$. By taking a cube root $v$ is obtained, and then $x = u - v$ can be found. The answer obtained is

$$x = \sqrt[3]{\sqrt{108} + 10} - \sqrt[3]{\sqrt{108} - 10}$$

You can check that $x = 2$ satisfies this equation. In fact, the above expression is equal to 2; the method gives the answer in an unusual form.

Even stranger, real roots are sometimes given in terms of complex numbers. The practice of expressing roots in this way was very perplexing to the Italian algebraists. Before this time square roots of negative numbers could safely be ignored as meaningless. Mathematicians began to think that it would be to their advantage to make sense out of negative square roots since they often did appear. For example, given

$$x^3 = 15x + 4$$

a root is

$$x = 4$$

but the rule for this case gives, as was shown later,

$$x = \sqrt[3]{2 + \sqrt{-121}} + \sqrt[3]{2 - \sqrt{-121}}$$

If you have studied complex numbers, you can show that this is, in fact, four, but it certainly seems to be a strange representation.

Cardano also encountered complex numbers in the problem of dividing 10 into two parts such that their product is 40, i.e., $x + y = 10$, $xy = 40$. By applying the rules, one gets $x = 5 + \sqrt{-15}$, $y = 5 - \sqrt{-15}$. Formally, it checks, but Cardano said this solution was "as subtle as it is useless." Cardano rejected negative roots as fictitious, probably taking for granted that the unknown represented a positive quantity.

Cubics and quartics had been solved, but general solutions had not been determined for higher degree equations. It became a famous problem to try to find a general solution to the fifth degree equation. In the 1800s this problem was shown to be impossible.

## ALGEBRAIC APPLICATIONS TO GEOMETRY

Algebra was developing during the Renaissance, and, as in the older Babylonian-Egyptian tradition, the application of numbers and algebra to geometrical problems was growing also. For example, **Bombelli (1560)** wrote an algebra book in which he solved the problem of finding a square inscribed in a given triangle algebraically. (His book was never published.) These algebraic applications to geometry were soon to increase in number.

Due to increased trade and exploration, there was a need for good maps. **Gerard Mercator (1596)**, a German whose real name was Gerhard Kremer, developed a new means of projecting a sphere onto a plane, called the *Mercator projection*, which has proved quite useful to this day in mapmaking, particularly since a fixed compass course can be represented by a straight line. Mercator developed his projection empirically, but the mathematical equations representing the projection were given about 30 years later.

**Francesco Maurolico** and **Federigo Commandino** translated the works of Archimedes, Apollonius, and Pappus at about 1550–60, and they solved some problems on centers of gravity in the Archimedean manner. By the end of the Renaissance scholars were able to appreciate the ancient works, most of which were available by that time.

# 3 Toward Modern Mathematics

Most of the Greek works we have now were already recovered and translated by the early 1600s. There were capable men in the new algebraic tradition who understood these works and added to them. These men had in their possession the algebra based on Arabic sources which had been developing in Europe for many years. It was a combination of the developing algebra and the advanced Greek geometry which led to a very fruitful period of discovery and to new powerful mathematical techniques. The seventeenth century was quite an exciting period for mathematics.

## ALGEBRA AS AN ANALYTIC ART

**Francois Viète (1580)** of France was a lawyer and a member of parlement. He used the method of Archimedes with inscribed polygons to approximate $\pi$, using polygons of $6 \cdot (2^{16}) = 393,216$ sides to find $\pi$ to nine decimal places. Soon after, **Adriaen van Rooman** in 1593 used polygons of $2^{30}$ sides to calculate $\pi$ to 15 places. In 1596 **Ludolph van Ceulen** used polygons of $2^{62}$ sides to compute $\pi$ to 35 places. He spent many years on this computation and had the result engraved on his tombstone.

Viète was among the first mathematicians to express algebra as a symbolic activity. That is, the unknown $x$ in an equation is merely a symbol. Algebra, or as Viète called it, *the analytic art*, is the art of manipulating symbols, such as changing $3x^2 - 2x = 4x + 1$ to $3x^2 = 6x + 1$. He believed this art to be a very powerful method of finding truth. At the end of his book Viète writes as his motto, *To leave no problem unsolved*.

One specific innovation of Viète's was to use consonants for given fixed but unspecified quantities and to use vowels for unknown quantities. Thus, finally a general equation such as $ax^2 = bx + c$ could be written, although Viète's symbolism did not produce an equation exactly like this. Before Viète, whenever a method was described, it was done using a given numerical example, such as $6x^2 = 2x + 3$. The distinction between a parameter $b$ and an unknown $x$ is generally a hard one for students to understand, so it is not surprising that parameters were developed rather late in history.

Viète's analytic art can be used to analyze various problems in algebra, in geometry, and in other fields. It was a significant development toward a more abstract mathematics. For example, he showed that the construction of a regular heptagon leads to an equation $x^3 = ax + a$. Here $x$ represents the side. This type of analysis, applied to geometry, soon led to the development of analytic geometry.

## AN ADVOCATE OF DECIMALS

**Simon Stevin (1585),** a Flemish engineer, was a very novel thinker. Finally, after all the years that the Hindu-Arabic numerals had been used in Europe, Stevin recognized the best way to write fractions using these numerals. Stevin wrote a book, *The Tenth*, advocating the use of decimal fractions and showing how to compute with them. The great advantage of decimals is that multiplication and division are done in the same manner as for whole numbers, only the decimal point has to be placed correctly in the result. Considering any other scheme of fractions, we find decimals are by far the best. Stevin's notation for a decimal fraction is illustrated by the following:

$$2 \; ⓪ \; 7 \; ① \; 6 \; ② \; 3 \; ③ \; 8 \; ④ = 2.7638$$

(The decimal point notation was first used a few years later by Napier.) Stevin also advocated that units of weights and measures be changed to decimal multiples, but this was not done in France until the Revolution in 1789, in England until relatively recently, and the gradual change is being made in the United States.

Stevin made some center of gravity calculations in a manner somewhat similar to that of Archimedes. For example, he showed that the center of gravity of a triangle lies on its median (fig. 4.4). He drew parallelograms with sides parallel to the median. By symmetry, the center of gravity of each parallelogram is on the median. As the triangle is divided into more and more parallelograms, the parallelograms more closely approach the triangle. Since the figure defined by the parallelograms always has its center of gravity on the median, Stevin reasoned that the triangle must also have its center of gravity on the median.

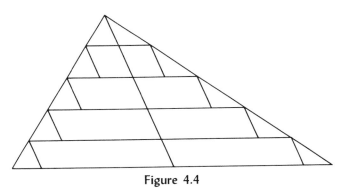

Figure 4.4

Stevin and other mathematicians of the time were developing methods for solving various problems which would now be solved by the general methods of calculus. In fact, it was the work of these men during the next 100 years which led to the creation of the calculus.

## THE DISCOVERY OF LOGARITHMS

**John Napier (1600),** a Scot, made a set of multiplication tables on sticks, which were called *Napier's rods* (fig. 4.5). They were used for gelosia, or grating multiplication. Thus, one could use Napier's rods to multiply without using the multiplication tables. To multiply 25 × 34 one would use the 2-stick and the 5-stick for 25 (fig. 4.5a). Using the third and fourth rows for 34 (fig. 4.5b) for the multiplication, one obtains the numbers needed without having to memorize multiplication tables.

Napier was an eccentric so many anecdotes about him have survived. One story states that Napier became extremely irritated, because his neighbor's birds continually flew over his land. Napier told his neighbor

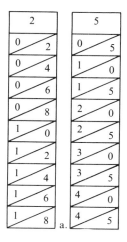
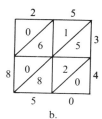

**Figure 4.5**

to keep the birds away, but the neighbor refused to cooperate, saying that if Napier could catch the birds, he could keep them. Napier immediately set out some corn which had been soaked in whiskey. The birds ate the corn and wobbled about while Napier easily collected them in a sack. He was thought by some to have enchanted the birds.

Another tale concerns a problem Napier had with a servant who was stealing from him. He devised an interesting way to determine which servant it was — he claimed to have a psychic chicken. His servants were each told to touch the chicken so that it would tell him who was guilty. Napier had rubbed the chicken with coal so that when the servants touched it their hands would become black. He then ordered the servants to turn their hands over for examination. The servant with clean hands was shown to be the guilty party, since he did not touch the chicken for fear of being revealed.

Napier lived for a time in an old house near a mill. The clack of the mill would disturb his work, so he would go out in his nightgown and cap to ask the miller to stop grinding. He probably appeared quite odd, and, in fact, he was considered to be a warlock by some local folk. The people believed that Napier had a compact with the devil and that the time he spent in study was spent in learning the black art.[6]

Napier's most important contribution to mathematics was to develop a table of logarithms. Logarithms helped to simplify calculations in astronomy which were very laborious, often involving multiplication and division of numbers to 10 or 15 decimal places. Using logarithms, multiplication and division can be replaced by addition and subtraction. (A

---

[6]Mark Napier, *Memoirs of John Napier of Merchiston* (Edinburgh: William Blackwood, 1884), p. 215.

slide rule is nothing more than logarithms on a stick.) For example, suppose that in the relation $x/A = B/C$, the numbers $A$, $B$, and $C$ are given, and we wish to find $x$. Rather than multiplying and dividing, we use a log table to find log $A$, log $B$, and log $C$. Then we have log $x$ − log $A$ = log $B$ − log $C$, so we can find log $x$ and then look up $x$ in the table. If we have a log table, we need only add and subtract to find $x$.

Logarithms to the base 10 are most familiar to us, but Napier did not use the idea of a base. He started with a circle of radius 10,000,000. In trigonometry we now use a circle of radius 1, but before decimals it was customary to use a large radius, such as $10^7$, so that sines would come out to be integers, rather than fractions. Even though Napier was familiar with decimals, he continued the custom of using the large value for the radius.

Napier wanted to get a reasonable number of points along the radius at which he could find the logarithms. Consider in base 10 the points with logarithms which are easy to find — $10^1$, $10^2$, $10^3$, $10^4$, $10^5$, . . . , $10^n$, which have logarithms of 1, 2, 3, 4, 5, . . . , $n$, respectively. These points are in a geometric progression increasing by 10 times at each step. Napier chose his method because it allowed him to find points reasonably close together which were in geometric progression. The logarithms of these points were then easy to determine. The following is Napier's definition of a logarithm.

**Definition** Let a line segment $TS$ (a radius of length 10,000,000) and a half line, or ray, $BI$ be given (fig. 4.6). Let a point $P$ start from $T$ and move along $TS$ with variable speed decreasing in proportion to its distance from $S$; at the same time let a point $Q$ start from $B$ and move along $BI$ with uniform speed equal to the rate with which $P$ began its motion. The variable distance $BQ$ is the logarithm of the distance $PS$.

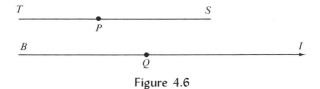

Figure 4.6

It is perhaps difficult to see that Napier's concept bears any relation to logarithms as we know them. In his system log 10,000,000 = 0, since at the start $P$ is at $T$, and $Q$ is at $B$. Thus, $PS = 10,000,000$, and $BQ =$ log $PS = 0$. Also, Napier's logarithm, $BQ$, increases as the number, $PS$, decreases, in contrast to the base 10 system in which the logarithm increases as the number increases. Nevertheless, despite these differences, it is true in Napier's system that if $x/A = B/C$, then $\log x - \log A = \log B - \log C$, so that $x$ may be found using Napier's tables.

We will illustrate that the above property is true by means of an example.

**Example** Consider the numbers (in millions) $x = 8$, $A = 4$, $B = 2$, and $C = 1$ (fig. 4.7). It is true that $8/4 = 2/1$.

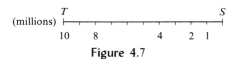

Figure 4.7

Since the speed of $P$ at any point is proportional to its distance from $S$, the speed of $P$ at 8 is $8/10$ of its speed at the start $T$, and the speed of $P$ at 2 is $2/10$ of its speed at the start $T$. Thus, the speed of $P$ at 8 is four times the speed of $P$ at 2. Also, $P$ travels four times as far to get from 8 to 4 as it does to get from 2 to 1 on the line. Since $P$ is going four times as fast from 8 to 4 and four times as far, it takes exactly the same time to go from 8 to 4 as it does to go from 2 to 1.

Now, consider the point $Q$ on the logarithm ray $BI$ (fig. 4.8). The point $Q$ is moving at constant speed.

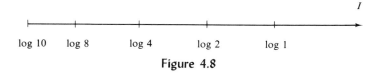

Figure 4.8

When $P$ is at 8 on the line $PS$, $Q$ is at log 8 on the ray $BI$. When $P$ is at 4, $Q$ is at log 4. Thus, in the time it takes $P$ to get from 8 to 4, $Q$ moves from log 8 to log 4. Similarly, in the time it takes $P$ to get from 2 to 1, $Q$ moves from log 2 to log 1. Since it takes $P$ the same time to get from 8 to 4 as it does from 2 to 1, it takes $Q$ the same time to get from log 8 to log 4 as it does to get from log 2 to log 1. Since $Q$ is moving at constant speed, it covers the same distance from log 8 to log 4 as it does from log 2 to log 1. Thus,

$$\log 8 - \log 4 = \log 2 - \log 1$$

Analogous reasoning can be used to show, in general, that if $x/A = B/C$, then

$$\log x - \log A = \log B - \log C$$

Of course, in order to make use of this general formula to simplify calculations, Napier needed a table of logarithms. The following illustrates how Napier made his table. He first found upper and lower bounds for log 9,999,999. To find the lower bound, consider the point $P$ to have moved one unit along $TS$ to point $R$, so that $RS = 9,999,999$ (fig. 4.9).

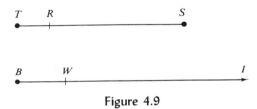

Figure 4.9

In the same time, the point $Q$ will move more than one unit since it is moving at a constant speed, while the speed of $P$ is decreasing. Thus, $TR < BW$. By definition, $BW$ is the log of 9,999,999, so that

$$1 < \log 9{,}999{,}999$$

To obtain an upper bound for log 9,999,999, Napier had the clever idea of letting the point $P$ start moving even before it gets to $T$. He defined a new point $O$ such that $P$ takes the same time to get from $O$ to $T$ as it did to get from $T$ to $R$ (fig. 4.10). Since the speed of $P$ is proportioned to its distance from $S$, in the same time it will cover the same fraction of its length, no matter where it starts. Thus

$$\frac{OS}{OT} = \frac{TS}{TR}$$

Figure 4.10

and by algebra,

$$\frac{OS - OT}{OT} = \frac{TS - TR}{TR}$$

giving

$$\frac{TS}{OT} = \frac{RS}{TR}$$

or

$$OT = \frac{TS \cdot TR}{RS}$$

In this instance, $TS = 10{,}000{,}000$ while $TR = 1$ and $RS = 9{,}999{,}999$, so

$$OT = \frac{10{,}000{,}000}{9{,}999{,}999} \approx 1.0000001$$

Napier defined the point $O$ so that in the same time that $P$ went from $O$ to $T$, $Q$ went from $B$ to $W$. But $P$ was always going faster than, or as fast as, $Q$ since its final slowest speed at $T$ equals the constant speed of $Q$ at $B$ and along its motion. Since $P$ is going faster, it covers more distance in the same time and

$$OT > BW$$

or

$$1.0000001 > \log 9{,}999{,}999$$

Napier took as his value for $\log 9{,}999{,}999$ the average of the upper and lower bounds, so that

$$\log 9{,}999{,}999 = 1.00000005$$

a figure accurate to seven places. He could have achieved greater accuracy by letting $P$ move for a shorter time interval so that the upper and lower bounds, $OT$ and $TR$, would be closer together. Similarly, if Napier had let $P$ move to $9{,}999{,}000$, he would have achieved less accuracy.

It was now easy for Napier to obtain the logarithms of certain other numbers. He just let $P$ move for the same time it took to get from $T$ to $R$. In this time, $P$ gets from $R$ to $R_1$ (fig. 4.11). Since $TR$ is $1/10^7$ of $TS$,

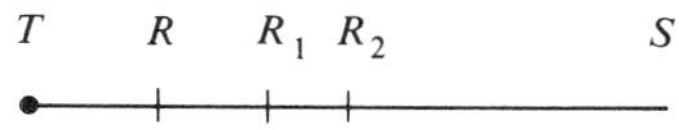

Figure 4.11

$RR_1$ is $1/10^7$ of $RS$, or

$$\frac{1}{10^7}(9{,}999{,}999) = .9999999$$

Thus,

$$R_1S = 9{,}999{,}999 - .9999999 = 9{,}999{,}998.0000001$$

The point $Q$ takes the same time to go from $B$ to $W$ as it does to go from $W$ to $W_1$ (fig. 4.12), and since it is going at constant speed, $WW_1 = BW = 1.00000005$. Thus,

$$\log R_1S = \log 9{,}999{,}998.0000001 = 2.00000010$$

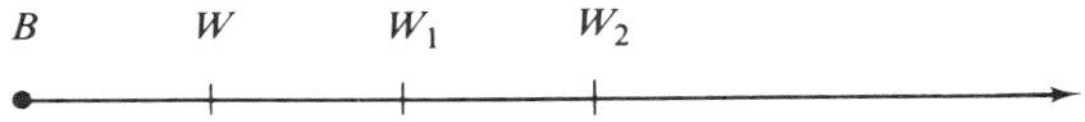

Figure 4.12

Napier repeated the process. The line $R_1R_2 = 1/10^7\ R_1S \approx .9999998$, giving

$$R_2S = 9{,}999{,}998.0000001 - .9999998 = 9{,}999{,}997.0000003$$

Since $W_1W_2 = WW_1 = BW$,

$$\log R_2S = 3.00000015$$

In a similar manner, using only subtraction and addition, Napier could compute the logarithms of as many numbers as he chose in this sequence. Of course, continuing in steps of such small size, he would never finish the calculations. In any case, he did not want a table with over 10,000,000 entries, so he continued for only 100 intervals, (*little steps*), at which point he found the logarithm of 9,999,900.0004950. He then used a method (which we will not describe) to interpolate and obtain the logarithm of 9,999,900.0000000.

Napier defined a new time interval, namely, that time necessary for $P$ to get from 10,000,000 to 9,999,900.0000000. He proceeded to find the logarithms of the sequence of distances that $P$ would reach in 50 such successive time intervals. The first distance would be

$$\begin{array}{r} 9{,}999{,}900.0000000 \\ 99.9990000 \\ \hline 9{,}999{,}800.0010000 \end{array}$$

and its log would be twice the log of 9,999,900.0. The fiftieth such *medium step* got Napier to 9,995,001.224804. Again, the steps proved too small, so Napier found log 9,995,000.000000.

Napier then let $P$ move for 21 large time intervals, each equal to the time needed to get from 10,000,000 to 9,995,000. He was easily able to find the logarithms of those 21 lengths, the last length being 9,900,473.57808. The steps still were too small, so Napier took 69 *giant steps*, each in the same time it took to get from $10^7$ to 9,900,000. By means of these 69 giant steps $P$ moved all the way to 5,048,858.8900, so that Napier was able to complete his table. The little steps and the medium steps were only necessary to obtain sufficient accuracy. His table actually consisted of the 69 giant steps with 21 large steps between each giant step. Thus, the table had $69 \times 21 = 1449$ entries. He used it to make a log sin table. Napier also gave a method for finding the logarithms of numbers between 0 and 5,000,000, as well as a method for interpolating between table values.

One of the advantages of base 10 logarithms is that they are well suited to the decimal system. Only a table of logs from 1 to 10 is needed to find all logarithms easily. For example,

$$\log 435.7819 = \log (4.357819)(10^2)$$
$$= \log 4.357819 + 2 \log 10$$

In base 10, log 10 is 1, so we have

$$\log 435.7819 = \log 4.357819 + 2$$

The log 4.357819 is the *mantissa* and the 2 is the *characteristic*, and the log is found easily by using a table from 1 to 10.

Napier discussed the advantages of base 10 with **Henry Briggs (1615)**, and Briggs developed a base 10 logarithm table. There are various methods that were used to calculate base 10 logs, the following example being an interesting way of finding log 2.

*Method*   The log of any number can be estimated just by counting the number of its decimal digits. For example,

$$4 < \log 93647 < 5$$

since

$$10^4 < 93647 < 10^5$$

Now, $2^{1000}$ has 302 decimal places. Therefore,

$$301 < \log 2^{1000} < 302$$

and

$$301 < 1000 \log 2 < 302$$

and

$$.301 < \log 2 < .302$$

To obtain more accuracy, note that the number $2^{100,000,000}$ has 30,103,000 decimal places, so

$$.30102999 < \log 2 < .30103000$$

It is not necessary to multiply out $2^{1000}$ or $2^{100,000,000}$ and count the decimal places. Since all that is needed is the number of decimal places, the multiplication can be rather sloppy, just carrying along the first five figures. Note that the square of a number has either double the number of decimal places or has one less than double the number of decimal places. In the following table, only the first five figures of the number are retained, starting with $2^{20}$.[7]

| $n$ | $2^n$ | Number of decimal places in $2^n$ |
|---|---|---|
| 1 | 2 | 1 |
| 2 | 4 | 1 |
| 4 | 16 | 2 |
| 8 | 256 | 3 |
| 10 | 1024 | 4 |
| 20 | 10486 . . . | 7 |
| 40 | 10995 . . . | 13 |
| 80 | 12089 . . . | 25 |
| 100 | 12676 . . . | 31 |
| 200 | 16069 . . . | 61 |
| 400 | 25823 . . . | 121 |
| 800 | 66680 . . . | 241 |
| 1000 | 10715 . . . | 302 |

[7] For the complete table up to $n = 100,000,000$, see John Napier, *The Construction of Logarithms*, pp. 99–100.

Toward Modern Mathematics 149

## ASTRONOMER-MATHEMATICIANS

**Johann Kepler (1610),** a German, liked Copernicus' theory of planetary motion because it gave the central place in the universe to the sun. Kepler, in fact, verged on being a sun-worshipper. He wrote,

> In the first place, lest perchance a blind man might deny it to you, of all the bodies in the universe the most excellent is the sun, whose whole essence is nothing else than the purest light, than which there is no greater star; which singly and alone is the producer, conserver, and warmer of all things; it is a fountain of light, rich in fruitful heat, . . . .[8]

Kepler searched for mathematical harmonies in the motion of planets. He believed that there were six planets (the planets then known were Mercury, Venus, Earth, Mars, Jupiter, and Saturn) because 6 was a perfect number, and that God placed them at their distances because between the spherical orbit of one planet and the spherical orbit of the next could just be fit one of the five regular polyhedra. How nice! In the orbital sphere of Saturn Kepler inscribed a cube. In the cube he inscribed the orbital sphere of Jupiter, while in that sphere he inscribed a tetrahedron, and so on.

Kepler was associated with a great astronomer, Tycho Brahe, and inherited his extensive data. He spent years looking for harmonies representing the music of the spheres, and he found many which he expressed, some in musical notation. Kepler insisted that his harmonies and laws fit the available data. He was quite firm about this and also bold, in that he broke with tradition, giving up the circular uniform motion which had been assumed for centuries. The following are Kepler's three famous laws of planetary motion which later so influenced Isaac Newton.

1. Planets move in ellipses with the sun at one of the foci.
2. The vector from the sun to the planet sweeps out equal areas in equal times (fig. 4.13).
3. The squares of the periods of revolution are to each other as the cubes of the semimajor axes (fig. 4.14).

Imagine the amount of computation Kepler must have done in order to discover the last law. He impatiently awaited the development of logarithms and actually had begun to work on them himself.

---

[8]E. A. Burtt, *The Metaphysical Foundations of Modern Science*, p. 59.

$T$ = time to complete one revolution

**Figure 4.13**     **Figure 4.14**

It is significant for the evolution of mathematical science that Kepler believed "perfect knowledge is always mathematical." The founders of modern astronomy and physics believed in mathematics as *truth*. In the twentieth century we tend to think of mathematics more as a language and of science as providing only an analogy to the real world, yet we still carry forth the tendency of the Pythagoreans and Plato to value the abstract form over the concrete sense experience. For example, we pound on a hard solid table, but then claim that this table is really full of holes. We say it is made up of atoms with electrons and nuclei, between which there is mostly empty space. In the 1600s men of science believed in mathematics as truth. This inspired them to create physical laws from which our modern science and technology evolved. We now do not believe that science encompasses all reality, because it does relegate sense experience to secondary importance.

Kepler also made contributions of a more strictly mathematical nature, rather than only contributions in the field of astronomy. He wanted to make sure that wine merchants were not cheating him, so he computed the volumes of various shapes of barrels. In doing this he divided figures into sums of lines (fig. 4.15), developing techniques for finding the volume which avoided Archimedean type proofs. Kepler felt that just as the eye is made to see color, the mind is made to understand quantity.

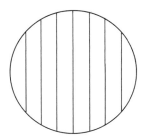

**Figure 4.15**

**Toward Modern Mathematics**

**Galileo Galilei (1630),** an Italian, had beliefs about mathematics similar to those of Kepler. He wrote,

> Philosophy is written in that great book which ever lies before our eyes — I mean the universe — but we cannot understand it if we do not first learn the language and grasp the symbols in which it is written. The book is written in mathematical language, and the symbols are triangles, circles, and other geometrical figures, without whose help it is impossible to comprehend a single word of it; without which one wanders in vain through a dark labyrinth.[9]

Galileo was familiar with the medieval work of Oresme, and others, on motion. He developed the basic concepts of motion, studying motion on an inclined plane, free fall, showing how velocity changes, and discussing acceleration. He also discovered that the path of a projectile is a parabola. Galileo's studies of motion greatly influenced Newton's study of gravitation.

Galileo discussed the infinite, mentioning the paradox of the squares — 1, 4, 9, 16, 25, . . . — corresponding in a one-to-one fashion to the integers — 1, 2, 3, 4, 5, . . . . Thus, in this sense, there are just as many squares as integers, yet the squares are a proper subset of the integers. Infinite sets are peculiar compared to finite sets.

Galileo also developed a sector compass and explained various uses for it, such as dividing a line segment into any number of parts, finding a fraction (such as 113/19) of a line, and increasing or decreasing the scale of a drawing.

**Bonaventura Cavalieri (1630),** another Italian mathematician, was a student of Galileo. He followed Kepler and Galileo in regarding areas as being made up of indivisibles, or lines, and worked quite hard to find the area under $y = x^n$, a problem we now calculate as $\int_0^a x^n \, dx = (a^{n+1})/(n + 1)$. Cavalieri started by considering a parallelogram divided into two triangles by its diagonal (fig. 4.16). He supposed that the parallelogram could be made up of lines parallel to its base. He then showed that

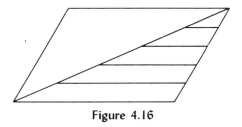

**Figure 4.16**

---

[9]Ibid., p. 75.

the sum of the lines in one triangle equalled 1/2 the sum of the lines of the parallelogram, while the sum of the squares of the lines of the triangle equalled 1/3 the sum of the squares of the lines of the parallelogram. These results were obtained by involved reasoning using the binomial expansion. Based on the results, Cavalieri could find the sum of the lines under a graph such as $y = x^2$, and thus find its area. With much more effort, he proceeded to solve the problem for higher powers of $x$.

Cavalieri is one of several men of his time who were developing methods for finding areas or for finding tangents to curves. After studying their involved methods one can appreciate the power of the calculus which makes many of those same problems seem easy in comparison.

Cavalieri's approach, the dividing of figures into infinitely many thin lines, was eventually abandoned because it was difficult to understand the correct way to use it. Infinite sets had often been considered paradoxical; Eudoxus and Archimedes had avoided dividing figures into infinitely many indivisibles for this reason. Cavalieri also discussed such a paradox. Consider a nonisosceles triangle (fig. 4.17). The perpendicular from the vertex divides the triangle into two right triangles. Each indivisible in the left triangle can be paired with an equal indivisible in the right triangle; therefore, the two triangles are equal in area! Obviously, something is wrong in this approach. Cavalieri was careful not to use indivisibles in this way.

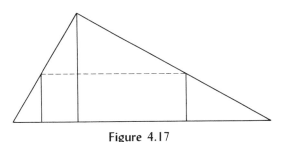

Figure 4.17

To obtain another paradox, consider concentric circles with the radius of the larger being twice the radius of the smaller (fig. 4.18). Regard each circle as made up of radii fanning out from the center. Each indivisible of the larger circle is twice that of the corresponding indivisible of the smaller circle, so, therefore, the area of the larger circle is twice that of the smaller circle! Unfortunately, the area of the larger circle is in fact four times that of the smaller.

In this last period from 1575 to 1630 the level of mathematical achievement was rising, lighting the way for new breakthroughs in both mathematics and science. Viète's algebra led to analytic geometry, while area, volume, and center of gravity calculations led to calculus. Newton's

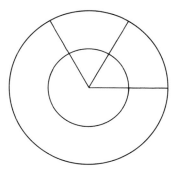

Figure 4.18

great theory of motion owed much to the results of Galileo and Kepler. Generally in mathematics, when new ideas shine, they do so in the light of the past.

## Problems

1. In finger reckoning the number of tens was shown using the thumb and index finger of the left hand, as indicated in figure 4.19. The numbers 100–900 were made using the same positions as those for 10–90, but with right hand, while the numbers 1000–9000 were made using the same positions as those for 1–9, but with the right hand. (Numbers from 10,000 to 1,000,000 were represented by placing the fingers in various positions on the body.) Form the following numbers on your fingers.

   a. 45   b. 72   c. 37   d. 850   e. 903
   f. 284  g. 519  h. 3728 i. 1461  j. 5376

10    20    30    40

50    60    70    80    90

Figure 4.19

2. Finger reckoning was also used to avoid memorizing multiplication tables from 5 × 5 to 9 × 9. Consider 6 × 8, for example. On the left hand, bend over the number of fingers, one, by which 6 is in excess of 5. On the right hand, bend over three fingers, representing the excess of 8 over 5.
   a. Add the number of bent fingers to get the tens digit of the product, and multiply the number of upright fingers on the left hand by the number of upright fingers on the right hand to get the units digit of the product.
   b. Find 7 × 9 in the same manner.
   c. Find 6 × 6. [*Note:* You will have to carry as the units are more than ten.]

3. Products from 10 × 10 to 15 × 15 can also be found by a method similar to that of problem 2.
   a. Find 12 × 14. Represent on the left hand the excess of 12 over 10, and on the right hand the excess of 14 over 10. The sum of the bent fingers represents the tens digit while the product of the bent fingers represents the units digit. Add 100 to obtain the final answer.
   b. Find 13 × 14 in the same manner.
   c. Find 13 × 13.

4. Do the following using Gerbert's method of division.
   a. 652 ÷ 8
   b. 437 ÷ 6
   c. 5721 ÷ 38 (use 40 − 2)
   d. 7945 ÷ 27

5. Gerbert actually developed his method of division for use on a line abacus. Instead of using seven counters to represent 7, he would use one counter with the numeral 7 marked on it, and so on for other numbers from 1 to 9. However, his division will work also on the traditional line abacus.
   a. Make a set of counters like Gerbert's by cutting cardboard and marking the pieces with numerals from 1 to 9. You will need several of each. Solve problem 4 using these counters on a line abacus.
   b. Using checkers or other markers as counters, solve problem 4 on a line abacus. (A table will serve as an abacus.)

6. Use the borrowing and repaying plan to find the following, writing each step in words.
   a.    3621
       − 1748
   b.    7238
       − 3875

7. Find the first 25 Fibonacci numbers.

8. Prove the following identity for Fibonacci numbers.

$$u_{n+m} = u_{n-1}u_m + u_n u_{m+1} \quad \text{for all } m \geq 1, n > 1$$

[*Hint:* Use induction on $m$. Show that the identity is true for $m = 1$. Then assume it true for $m = k$ and $m = k + 1$, and prove it true for $m = k + 2$.]

9. Prove that if $u_n$ and $u_m$ are Fibonacci numbers and $n$ is divisible by $m$, then $u_n$ is divisible by $u_m$. [*Hint:* Let $n = mm_1$ and prove by induction on $m_1$. Use the identity in problem 8.]

10. Use problem 9 to prove that
    a. if $n$ is divisible by 5, then $u_n$ is divisible by 5.
    b. if $n$ is divisible by 8, then $u_n$ is divisible by 7.

11. Perform the following divisions using the method of Planudes.
    a. $147 \div 13$          b. $900 \div 25$

12. Add the following numbers using the method of Sacrobosco.
    a. $437 + 524$    b. $6345 + 4731$    c. $3801 + 589$

13. Perform the following multiplications using the method of *The Crafte of Nombrynge*.
    a. $37 \times 682$         b. $49 \times 473$
    c. $376 \times 5812$     d. $63 \times 64821$

14. Perform the following multiplications using the Russian peasant method.
    a. $14 \times 29$       b. $23 \times 34$
    c. $36 \times 48$       d. $62 \times 104$

15. Write the following expressions using the notation of Chuquet.
    a. $6x^4$        b. $7x^{-5}$        c. $3x^2 + 4x^{-3}$

16. Solve the following using cross multiplication.
    a. $34 \times 52$       b. $46 \times 28$

17. Explain how cross multiplication could be extended to solve problems such as $28 \times 367$.

18. Solve the following expressions for $x$ using the method of Cardano.
    a. $x^3 + 3x = 10$      b. $x^3 + 6x = 2$

19. Solve the following examples of a different type of cubic than in the previous problem. [*Hint:* Let $x = u + v$ and follow Cardano's method.]
    a. $x^3 = 6x + 40$      b. $x^3 = 6x + 6$

20. Write the following decimals using the notation of Stevin.
    a. $3.683$        b. $24.1049$        c. $576.38842$

21. Use Napier's method of finding bounds for log 9,999,999 to find lower and upper bounds for log 9,999,000. Compare the accuracy of the bounds with those Napier obtained for log 9,999,999.

22. After taking 100 little steps, Napier obtained the logarithm of 9,999,900.0004950. In order to start his medium steps he needed to find log 9,999,900.000000. He developed a procedure which he used for this and similar problems. Let $TS$ be the radius, $dS$ the greater of the two given numbers, and $eS$ the lesser of the two given numbers (fig. 4.20). Let $V$ be chosen so that $ST/VT = eS/de$, and $C$ be chosen so that $TS/TC = dS/de$.
    a. Add 1 to both sides of $TV/ST = de/eS$ to show that $VS/ST = dS/eS$.

**Figure 4.20**

b. Subtract 1 from both sides of $TS/TC = dS/de$ to show that $CS/TC = eS/de$.
c. From part b, $TC/CS = de/eS$. Add 1 to both sides to show that $TS/CS = dS/eS$.
d. Use part c to show that $\log CS = \log eS - \log dS$.
e. Show that $TC < \log CS < VT$. [*Hint:* Use parts a and c to show $VS/TS = TS/CS$. Then follow similar reasoning to that used by Napier in finding bounds for log 9,999,999.]
f. Use parts d, e, and the defining equations of $V$ and $C$ to show that a lower bound for $\log eS - \log dS$ is $TC = (TS \cdot de)/dS$ and an upper bound is $TV = (TS \cdot de)/eS$.
g. Apply part f to find upper and lower bounds for log 9,999,900.-0004950 − log 9,999,900.000000. Since Napier had already found that log 9,999,900.0004950 = 100.0000050, he could now find log 9,999,900.

23. Following Briggs, find a
    a. one-place approximation to log 3 by finding the number of decimal places in $3^{10}$.
    b. two-place approximation to log 3 from the number of places in $3^{100}$. [*Hint:* Retain only two significant figures after each multiplication, starting with $3^{10} \approx 60,000$].
    c. three-place approximation to log 3 from the number of decimal places in $3^{1000}$.

24. Another method Briggs used to find logarithms involved taking square roots. His method can be used in the following manner to find log 4. First,

$$\log \sqrt{10(1)} = \log 3.162 = \frac{1}{2} \log 10 = \frac{1}{2}(1) = .5000$$

Since 4 is between 3.162 and 10, find $\sqrt{10(3.162)} = 5.623$. Then

$$\log 5.623 = \log \sqrt{10(3.162)} = \frac{1}{2} \log 10(3.162)$$
$$= \frac{1}{2}(\log 10 + \log 3.162)$$
$$= \frac{1}{2}(1 + .5000) = .7500$$

Since 4 is between 3.162 and 5.623, continue by finding $\sqrt{(3.162)(5.623)} = 4.217$ so that

$$\log 4.217 = \log \sqrt{(3.162)(5.623)}$$
$$= \frac{1}{2}(\log 3.162 + \log 5.623)$$
$$= \frac{1}{2}(.5000 + .7500) = .6250$$

Continue in this way until log 4.00 is obtained. To achieve greater accuracy, more decimal places would have to be retained at each step.

25. Following the method of Cavalieri, consider a parallelogram (fig. 4.21) to be made up of infinitely many horizontal lines, one of which is $CB$. Let $BO = x$, $CO = y$, $DO = z$, and $SR = a$.

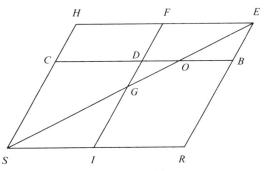

Figure 4.21

a. Show that $\sum a^2 = \sum (x + y)^2 = 2\sum x^2 + 2\sum xy$, where the summation is extended over all lines in the parallelogram.

b. Show that $\sum xy = 1/4 \sum a^2 - 1/4 \sum x^2$. [*Hint:* Let $x = a/2 - z$ and $y = a/2 + z$. Since $\triangle EGF$ is similar to $\triangle ESR$, each line in $\triangle EGF$ is 1/2 of a corresponding line in $\triangle ESR$. But $\triangle EGF$ is cut by only half as many lines as $\triangle ESR$, so

$$\sum_{\triangle EGF} z^2 = \frac{1}{2} \sum \left(\frac{x}{2}\right)^2 = \frac{1}{8} \sum x^2$$

and

$$\sum z^2 = \sum_{\triangle EGF} z^2 + \sum_{\triangle SGI} z^2 = 2 \sum_{\triangle EGF} z^2 = \frac{1}{4} \sum x^2$$

c. Use parts a and b to show that $\sum x^2 = 1/3 \sum a^2$, and, thus, that the sum of the squares of the lines in the triangle $ESR$ is equal to 1/3 the sum of the squares of the lines in the parallelogram.

## References

Europe in the Middle Ages

Boyer (1,2)
Grant
Haskins
Hofmann
Karpinski

Menninger
Smith (1)
Struik (3)
Vorobyov

The Renaissance

Boyer (1)
Cardano
Ivins
Karpinski
Kline (2)
Koestler

National Council of Teachers
  of Mathematics
Regiomontanus
Smith (1,2)
Struik (3)

Toward Modern Mathematics

Baron
Burtt
Cohen
Klein
Koestler

Napier
Smith (1,2)
Struik (3)
Toeplitz

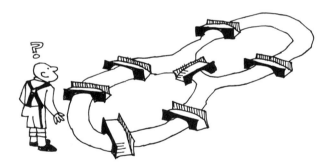

# 5 The Origin and Development of Analytic Geometry and the Calculus

We have seen in Viète's work in algebra the first glimmerings of mathematics as a symbol-manipulating activity. The seventeenth century emerged as a period in which the real power of the algebra that had been developing was displayed in applications to the study of curves, area and tangent problems, and motion. Analytic geometry and calculus are important creations of that century's mathematicians. Mathematical activity increased dramatically in the 1600s; number theory, probability, and projective geometry were developed. Yet the most significant advances were those in analytic geometry and the calculus, because with the tools they provided, problems that were impossible before could be solved quickly. Also, these new techniques were a magnificent aid in the creation of the new physical science. That science has been implemented in the technology which has influenced our present society so greatly.

## 1  Seventeenth Century Origin

### RENE DESCARTES (1596-1650)

A major philosopher, Descartes of France is responsible for the famous statement, "I think, therefore I am," the first principle of his philosophy. He cast aside beliefs based on authority and tradition, knowing that mistakes can be made in reasoning, and reasoned afresh his own conclusions. While considering established ideas to be false, Descartes realized that in the actual cognitive exercise of believing all to be false, he was aware of himself as a very real being. Thus, the basis of his philosophy was established.

Descartes was quite impressed with the clarity and preciseness of mathematical reasoning and hoped to use the mathematical method in developing his philosophy. He spent some years studying and writing about his method for distinguishing between truth and falsehood derived from mathematics. In 1637 he published a major work, *A Discourse on the Method of rightly conducting the Reason and seeking Truth in the Sciences. Further the Dioptric, Meteors, and Geometry, essays in this Method.* Descartes' main contribution to mathematics, *The Geometry*, was contained in his larger work as an example of how his method could be used to find new truths. *The Geometry* will be studied after a brief survey of Descartes' life.

Descartes discounted his formal education, reasoning that learning by experience was the more valuable process. As a result, he travelled outside of France, and even served in the military, finally settling in Holland for some years. Descartes was purportedly inspired to create his philosophy and mathematics by several dreams he had in 1619. It was fortunate for him that his family was wealthy, because he was able to devote his time to intellectual pursuits without worry of how to support himself. In Holland Descartes tutored the exiled Bohemian Princess Elisabeth, but most of his time was dedicated to contemplation and writing. He was later summoned to Sweden in 1646 to become tutor to Queen Christine. After many years of living very leisurely, Descartes found it impossible to adjust to the harsh Swedish winter and the rigorous schedule which Queen Christine has arranged. His health failed, and he died in 1650. Seventeen years after his death, Descartes' bones were returned to France, except those of his right hand which were kept as a souvenir by the official who arranged the transfer.[1]

Descartes was very proud of his general method in geometry. He wrote that the ancient Greeks proved only those theorems that they happened on, that they had no systematic method. He, however, used the analytic method in a systematic manner. This idea of *analysis*, assuming the problem is solved and working backward to something known, was known to the Greeks, but used sparingly. What made it powerful was the algebraic technique that Descartes used. In fact, he essentially applied to geometry the method of analysis which had previously been used in algebra. His work is a far-reaching development of

---

[1] See E. T. Bell, *Men of Mathematics*, pp. 35-55, for a biography of Descartes.

the analytic art of Viète. Descartes' analytic method of geometry is outlined in the four following steps, four steps that are the typical method of solving word problems in algebra. In the first two steps the word problem is translated into an equation. In step three the equation is solved, while step four is the check, or the *synthesis*, of the solution from the given information.

1. Assuming the given problem has a solution, introduce letters to represent known and unknown quantities.
2. Express the conditions of the problem as equations in the knowns and unknowns.
3. Use algebra to simplify and solve the equations. (To aid in this step Descartes made several contributions to the theory of equations, such as *Descartes' rule of signs* which will be described in problem 9 at the end of this chapter.)
4. Use the algebraic analysis to determine how to construct the solution. If the equation in part 3 was quadratic, then the solution could be constructed with a straightedge and compass. (Descartes also described an apparatus for constructing roots of higher degree equations, such as cubics, fourth powers, and so on.)

The following example is a problem from Descartes' book.

**Example** Given the square $ABDC$ and the line $BN$, prolong the side $AC$ to $E$, so that $EF$, laid off from $E$ on $EB$, shall be equal to $BN$ (fig. 5.1).

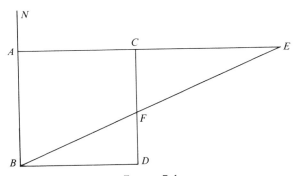

Figure 5.1

*Solution* Pappus had given a construction for the solution using the synthetic method of proceeding from the known to the unknown by construction. Descartes used the following analysis.

1. Consider the completed solution (fig. 5.1). Let $BD = a = CD$ and let $NB = c$, both of which are given. Let $DF = x$.
2. According to the condition of the problem, the line $EF$ equals $NB$, so $EF = c$. Since $CD = a$ and $DF = x$, we have $CF = a - x$. Triangles $CEF$ and $DBF$ are similar since angles $D$ and $C$ are right angles and $BFD$ and $CFE$ are vertical angles. Thus, the corresponding sides are proportional giving

$$\frac{CF}{FE} = \frac{FD}{BF}$$

or

$$\frac{a-x}{c} = \frac{x}{BF}$$

so

$$BF = \frac{cx}{a-x}$$

and by the Pythagorean theorem

$$(BF)^2 = (BD)^2 + (FD)^2$$

so

$$\left(\frac{cx}{a-x}\right)^2 = a^2 + x^2$$

Multiplying, we find that the unknown $x$ satisfies the equation

$$x^4 - 2ax^3 + 2a^2x^2 - 2a^3x + a^4 = c^2x^2$$

3. Descartes then used algebra to show that the root $x$, even though satisfying a quartic equation, could be expressed in terms of square roots only. The expression is

$$x = \frac{1}{2}a + \sqrt{\frac{1}{2}a^2 + \frac{1}{4}c^2} \\ - \sqrt{\frac{1}{4}c^2 - \frac{1}{2}a^2 + \frac{1}{2}a\sqrt{a^2 + c^2}}$$

4. Based on the above expression for $x$, a construction of $x$ using a straightedge and compass could be carried out starting from the given lines $a$ and $c$. In practice this was not done, but it was clear that it could be done, and Descartes felt that it was necessary to be able to construct the solution. Later it was considered to be sufficient to find the root algebraically, and step 4 has been deleted from the process.

Descartes was interested in developing a systematic method of obtaining theorems of Euclidean geometry. The above problem had a unique solution, but the influence of Descartes' method was not in solving this type of determinate problem. Descartes' greatest influence was felt in the area of the study of curves.

In *The Geometry* Descartes treated one problem extensively. This problem, called the *three- and four-line locus*, had been worked on by Euclid, Apollonius, and Pappus. Given four lines, the problem is to find the locus of points such that the product of the distances of the point to two of the lines (along specified directions) is equal to a constant times the product of the distances from the point to the other two lines (fig. 5.2). In figure 5.2 the four given lines are $AB$, $AD$, $EF$, and $GH$. The problem is to find all points $C$ such that $CB \cdot CF = kCD \cdot CH$, where the angles $CBE$, $CFE$, $CDA$, and $CHG$ are given in advance. Descartes solved the problem using his analytic method and taking as an example the case when $k = 1$. He let $AB = x$ and $CB = y$, and then expressed all

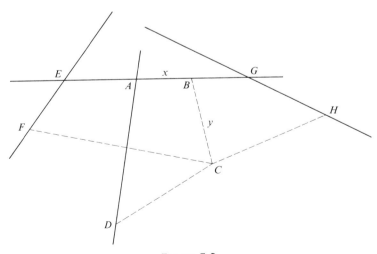

Figure 5.2

the other distances in terms of $x$ and $y$. The condition of the problem gave him the equation

$$y = m - \frac{n}{z}x + \sqrt{m^2 + ox + \frac{p}{m}x^2}$$

where $m$, $n$, $z$, $o$, and $p$, are defined by equations too involved to present here. This equation enabled Descartes to show that the locus of points was a conic section. He determined under what conditions the solution was a circle, a hyperbola, an ellipse, or a parabola. (Although Cartesian coordinates are justly named after Descartes, it would be misleading to think that he used coordinates in exactly the same way we do. His approach is represented in figure 5.2. The concept of two perpendicular axes was not common until the eighteenth century.)

Descartes did not stop with the three- and four-line locus problem. He went on to show that the solution to an analogous five- and six-line locus problem was a cubic curve. Thus, his method of analysis leads to the introduction and study of all types of new curves. This is an unattainable feat using Greek mathematics. Descartes' method was indeed powerful, because it made possible the algebraic deduction of properties of curves.

## PIERRE FERMAT (1601-1665)

Pierre Fermat, a Frenchman, was one of the greatest mathematicians of the 1600s, a man who contributed to and initiated several fields of mathematics. Mathematics was actually only a hobby of Fermat's, but he was able to spend a great deal of time on his hobby. Fermat was a judge, and as such he was expected to avoid society; if a judge remained aloof, he was not susceptible to bribery. Therefore, Fermat was able to devote most of his leisure time to mathematics.

In calculus, Fermat found tangents to curves and maxima and minima by methods essentially the same as our present ones. He was also able to find the area under various curves, with a method similar to that of Archimedes', but made much more powerful by the use of algebra. Fermat approximated the areas by finding the sum of the areas of rectangles under the curves (fig. 5.3). He found that the sum of the areas of the rectangles more closely approached the area under a curve as the number of rectangles increased, and using this approach, he was able to find the exact area under the curve.

Fermat developed the *principle of least time* in optics, a concept which proved to be influential in the twentieth century development of quantum mechanics. This principle states that the path of light from point $A$ to

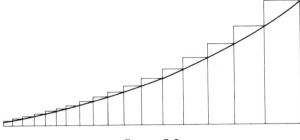

Figure 5.3

point $B$ as it passes through various media is the path which makes the time from $A$ to $B$ an extremum. From his principle of least time, Fermat could deduce the laws of reflection and refraction:

$$\text{angle of incidence} = \text{angle of reflection}$$
$$\text{sine of angle of incidence} = k \text{ sine of angle of refraction}$$

Fermat did his greatest work in the theory of numbers, the study of the properties of 1, 2, 3, 4, 5, . . . . Consider the following sequence of numbers called *Fermat numbers*.

| $2^{2^0}+1$ | $2^{2^1}+1$ | $2^{2^2}+1$ | $2^{2^3}+1$ | $2^{2^4}+1$ |
|---|---|---|---|---|
| 3 | 5 | 17 | 257 | 65,537 |

| $2^{2^5}+1$ | | $2^{2^6}+1$ | | . . . |
|---|---|---|---|---|
| 4,294,967,297 | | 18,446,744,073,709,551,617 | | |

The first five numbers are prime, but 641 divides $2^{2^5}+1$, and 274,177 divides $2^{2^6}+1$. Fermat believed that all the numbers were prime, but he did not claim to have proved that they were. No one has yet found any other Fermat numbers that are prime, but several numbers (including some very large ones) have been shown to be composite. It has not been proved, however, that no more Fermat primes exist. Fermat numbers are related to polygons constructible with a straightedge and compass, a fact which will be studied later.

Fermat made another discovery which is called *Fermat's little theorem*.

**Theorem** If $n$ is an integer, and $p$ is prime, then $p$ divides $n^p - n$.

**Example** 3 divides $1^3 - 1 = 0$, $2^3 - 2 = 6$, $3^3 - 3 = 24$, . . . .

Another theorem developed by Fermat and related to the previous theorem concerns perfect numbers.

**Seventeenth Century Origin** 169

**Theorem** If a prime $q$ divides $2^p - 1$, then $q = 2kp + 1$, where $k = 1, 2, 3, 4, \ldots$ and $p$ is an odd prime.

This last theorem greatly reduces the work involved in computing perfect numbers. Recall that if $2^p - 1$ is prime, then $2^{p-1}(2^p - 1)$ is perfect. For example, without Fermat's theorem, to find if $2^{29} - 1$ is prime, we would have to divide by all primes less than its square root; thus, we divide by 3, by 5, by 7, by 11, ..., and so on. With Fermat's theorem, we need only try primes of the form $2k(29) + 1$, for example,

$$2(1)29 + 1 = 59$$
$$\cancel{2(2)29 + 1 = 117} \quad \text{not prime}$$
$$\cancel{2(3)29 + 1 = 175} \quad \text{not prime}$$
$$2(4)29 + 1 = 233$$

Thus, the improvement over the older method is great, because one needs to try fewer numbers.

Fermat considered the odd primes to be of two classes — those which are one bigger than a multiple of 4, and those which are one smaller than a multiple of 4.

$$4n + 1 \quad \text{primes 5, 13, 17, \ldots}$$
$$4n - 1 \quad \text{primes 3, 7, 11, \ldots}$$

He claimed to have proved that every prime of the form $4n + 1$ is a sum of two squares and in only one way, apart from the order of the terms, and also that no number of the form $4n - 1$ is a sum of two squares. Thus,

$$5 = 1^2 + 2^2 \qquad 13 = 3^2 + 2^2 \qquad 17 = 4^2 + 1^2$$

Fermat read Diophantus and solved some problems similar to the Greek's. For example, he showed that $y = 3$, $x = 5$ is the only solution to $y^3 = x^2 + 2$, although he did not publish a proof. Fermat made notes in the margin as he read Diophantus. Next to the $x^2 + y^2 = a^2$ problem he wrote, "On the contrary, it is impossible to separate a cube into two cubes, a fourth power into two fourth powers, etc. I have discovered a truly marvelous demonstration which this margin is too narrow to contain." This conjecture is the so-called *Fermat's last theorem* which no one yet has been able to prove. In 1908 a German professor donated 100,000 marks to be awarded to the first person giving a proof, but inflation reduced this prize to a fraction of a cent.

Fermat also made contributions in the fields of analytic geometry and probability. Working independently of Descartes, Fermat developed analytic geometry, giving equations for curves, and together with Blaise Pascal, Fermat initiated the theory of probability, although Cardano had made an earlier contribution. Though great advances were made in the 1600s using analytic geometry, there was another interesting geometry being developed — the geometry of projections.

## GIRARD DESARGUES (1591-1661)

Desargues was a French engineer who wrote on projections with the goal of aiding painters and architects. In his book *Rough Draft of an Attempt to Deal with the Outcome of a Meeting of a Cone with a Plane* he introduced much unusual terminology. As a consequence, the text was not very popular, and all printed copies disappeared. In 1847 a handwritten copy was found, but Desargues' ideas had been rediscovered and published by others before that time. Another reason why his work was not popular is that analytic geometry and area problems were the celebrated innovations of the time. The new algebraic methods were powerful and popular; projective geometry was neglected.

Desargues studied the properties of projections of figures. If you have a light at a point and shine it on a circle, the projection on another plane at an angle to first plane might be an ellipse. It might also be a parabola or hyperbola. We can see this by thinking of a circle drawn on a pane of glass (fig. 5.4). Shine a light on the glass and note the shadow of that circle on another plane at an angle to the pane of glass.

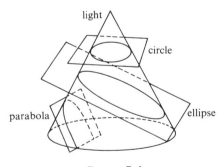

Figure 5.4

The shape or length of a curve can be changed by projection. Lines, however, project into lines, and intersecting lines will project into intersecting lines or parallel lines. Many statements of projective geometry would be shorter and simpler if it were not necessary to distinguish between intersecting lines and parallel lines. To achieve this simplification, parallel lines are by convention said to intersect in a point infinitely far away. Then, accepting this convention, we can say that every two lines intersect and that intersecting lines will always project into intersecting lines. The theorems that remain true under projection of a figure are the subject of projective geometry. The projective plane is the ordinary plane with points at infinity added.

An important theorem of projective geometry was found by Desargues and is called, therefore, *Desargues' theorem*.

**Theorem**  Consider a triangle $ABC$ and its projection $A'B'C'$ (fig. 5.5). If the corresponding extended sides $AC$ and $A'C'$ meet in a point $P$, while the extended sides $AB$ and $A'B'$ meet in $Q$, and the extended sides $BC$ and $B'C'$ meet in $R$, then $P$, $Q$, and $R$ lie on one straight line.

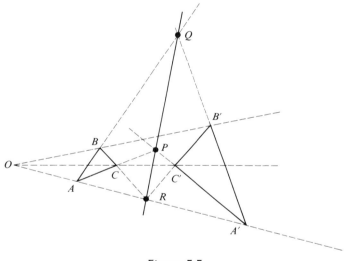

Figure 5.5

When a line segment is projected into another line segment its length generally changes. Desargues found a number that is invariant when one line is projected to another, called the cross-ratio. Given any four points, $A,B,C,D$, on a line, the *cross-ratio* can be defined as

$$\frac{CA/CB}{DA/DB}$$

The cross-ratio was known also to Pappus. Consider a line $l$ and its projection $l'$ (fig. 5.6). Take any four points on $l$, say $A$, $B$, $C$, and $D$,

172  The Origin and Development of Analytic Geometry and the Calculus

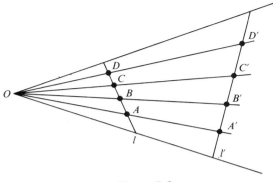

**Figure 5.6**

and let them be projected to points $A'$, $B'$, $C'$, and $D'$, respectively, on line $l'$. Although $A'B'$ is generally unequal to $AB$, and so on, the cross-ratio computed on $l$ is equal to the cross-ratio computed on $l'$,

$$\frac{CA/CB}{DA/DB} = \frac{C'A'/C'B'}{D'A'/D'B'}$$

### BLAISE PASCAL (1623-1662)

Pascal, a Frenchman and a student of Desargues, is recognized for his literary and religious writings, such as *Thoughts* and *Provincial Letters*. His father gave him lessons, but avoided teaching Pascal mathematics, because he thought it would be too much of a strain on the boy who was in poor health. Pascal's father was a mathematician himself and was pleased, therefore, when Pascal asked to learn about geometry at age 12. Pascal quickly demonstrated his mathematical skills when he proved by himself that the angles of a triangle add to 180°. His father was so happy at this accomplishment that he gave Pascal a copy of Euclid. At the age of 14 Pascal was admitted to the scientific discussions of Father Mersenne, who was a friend of Descartes and who also corresponded with the leading mathematicians of the time.

When he was almost 16 years of age, Pascal proved a theorem now known as *Pascal's theorem*.

**Theorem** Consider any conic section, say an ellipse. Take any six points on the ellipse — $A, B, C, D, E, F$ — and connect them by straight lines (fig. 5.7). The segments $AB$ and $DE$ are opposite sides, as are $BC$ and $EF$, and $CD$ and $FA$. The three points of intersection of the three pairs of opposite sides lie on one straight line.

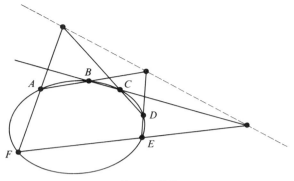

**Figure 5.7**

Pascal's theorem is an example of a theorem which remains true under projection. Pascal wrote a manuscript, *Essay on Conics*, containing 400 theorems and including many results of Apollonius and others as corollaries to his theorem obtained by moving the six points to various positions. This manuscript was never published and was ultimately lost, although two copies of a summary have survived.

While still quite young, Pascal made another notable contribution to mathematics. At the age of 18, he invented the adding machine, which has since evolved into very sophisticated calculators. With Pascal's machine, the operations of addition and multiplication (set up as repeated addition) could be performed.

Pascal and Fermat founded the theory of probability through correspondence (though Cardano had earlier done important work). A gambler had posed to Pascal the problem of how to divide the stakes when a dice game is interrupted in progress. Pascal had to find the probability of winning for each player. To count the possible cases Pascal and Fermat developed some combinatorial analysis. In this connection Pascal made use of a triangle used much earlier in China and even in the West.

```
                1
             1     1
          1     2     1
       1     3     3     1
    1     4     6     4     1
    .     .     .     .     .     .
    .     .     .     .     .     .
    .     .     .     .     .     .
```

As a simple illustration of a use of Pascal's triangle, consider row four — 1, 3, 3, 1. The sum of these numbers is eight. Suppose a coin is tossed three times (one less than the row number). The chances are one out of eight that three heads will occur, three out of eight that exactly two heads will occur, three out of eight that exactly one head will occur, and one out of eight that no heads will occur.

The following example is a modernized version of the type of problem Pascal and Fermat solved.

**Example**  A coin is tossed three times. $A$ and $B$ each bet $1, and the first to win two of the three tosses wins the bet. Suppose $A$ chooses heads and wins the first toss, but then has to leave. How should the bet be divided?

*Solution*  At the start of the game each person has an even chance of winning. Given that heads occurred on the first toss, there are four possibilities for the next two tosses. Of those four possibilities, only three are favorable to $A$, so his chances are 3/4 of winning if the game were to be continued.

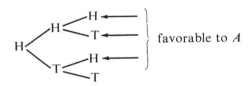

Thus, $A$ should get 3/4 of the $2 wager, or $1.50. $B$ should get 1/4, or $0.50.

Probability theory has since developed into an immensely useful subject. It is utilized in all the sciences, business, and engineering.

Pascal made some interesting discoveries concerning the cycloid, which was a popular curve for study during Pascal's time. The *cycloid* is the path described by a point on a rolling circle (fig. 5.8). Oddly enough, Pascal probably would not have made his discoveries if he had

Figure 5.8

not been suffering with a toothache; he worked on the cycloid to keep his mind off the pain. Pascal found the area of sections obtained by lines parallel to the base, and he found the center of gravity of each of these sections. These and other results were included in a book he wrote on the cycloid.

**Christiaan Huygens (1629–1695)** proved that the upside down cycloid is the *tautochrone*. Imagine a curve with beads placed at various points on the curve. If each bead takes the same amount of time to slide to the lowest point of the curve, then that curve is a tautochrone. Huygens also wrote a book on timekeeping in which he showed how to make a pendulum oscillate in a cycloid, so that each beat will take the same amount of time, by hanging it from cycloidal jaws (fig. 5.9). For practical purposes, however, Huygens found that a pendulum sweeping out a circular arc was sufficiently accurate.

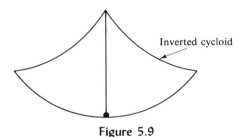

Figure 5.9

There were also other writers in the early middle 1600s who were developing methods for area problems, finding centers of gravity, and so on. These are the men whose work laid the foundation for the calculus of Newton and Leibniz. In addition to the mathematicians already mentioned are Gilles Persone de Roberval (1602–1675), Evangelista Torricelli (1608–1647), Gregory of St. Vincent (1584–1667), John Wallis (1616–1703), James Gregory (1638–1675), and Isaac Barrow (1630–1677).

## COMPUTATION OF $\pi$

**James Gregory** in 1673 found the series

$$\arctan x = x - \frac{1}{3}x^3 + \frac{1}{5}x^5 + \cdots$$

Substituting $x = 1$, then

$$\frac{\pi}{4} = 1 - \frac{1}{3} + \frac{1}{5} - \frac{1}{7} + \frac{1}{9} - \cdots$$

a very neat expression for $\pi$. However, this series converges too slowly to be of use in computing $\pi$. For example, the fiftieth term is 1/99, which is greater than .01.

In 1706 **John Machin** used Gregory's series for arctangent and the relation

$$\frac{\pi}{4} = 4 \arctan \frac{1}{5} - \arctan \frac{1}{239}$$

to compute $\pi$ to 100 decimal places. These series converge very rapidly. Consider, for example,

$$\arctan \frac{1}{5} = \frac{1}{5} - \frac{1}{3}\left(\frac{1}{5}\right)^3 + \frac{1}{5}\left(\frac{1}{5}\right)^5 - \cdots + \cdots$$

Here the second term is small and the third term is smaller still. Much greater accuracy can be obtained by using three terms of this series than by using 100 terms of the series for arctan 1.

**William Shanks** used Machin's expression to compute $\pi$ to 707 places, a task he completed in 1873 after 15 years of work. (Later, using computers, an error was found in the 528th place.) Shanks' computation of $\pi$ has even been printed in Ripley's *Believe It or Not*. The difficulty of this calculation is amazing, but with computers much better results have been obtained. The following table illustrates the progressively more accurate calculations of $\pi$ and the amount of time spent to make the calculations.

| Year | No. of places of $\pi$ | Time |
|---|---|---|
| 1949 | 2,037 | 70 hours |
| 1958 | 10,000 | 1 hr. 40 min. (The calculation of the first 707 places took 40 seconds) |
| 1961 | 100,265 | 8 hr. 43 min. |
| 1967 | 500,000 | 28 hr. 10 min. (The check required 16 hr. 35 min.) |

**Lindemann** in the 1880s proved that $\pi$ is transcendental, that is, $\pi$ is never the root of a polynomial equation. Knowing this fact, it can be shown that squaring the circle is impossible. Lindemann's proof was based on the proof of Hermite which proved $e$ transcendental.

## ISAAC NEWTON (1642-1727)

Isaac Newton of England made revolutionary advances in mathematics and physics, and he showed the potential of doing so at an early age. Being a frail child, Newton was more inclined to intellectual, rather than physical, exercise. He invented kites, mechanical toys, mills, and clocks while he was still a young boy. Because Newton showed such mental capacities, his uncle sent him to college.

Newton became very interested in mathematics while he was in college. He later attributed this interest to the fact that he was unable to understand a book on astrology because of the geometry and trigonometry in the text. Consequently, he studied Euclid's *Elements* and Descartes' *Geometry*. His mathematical interest kindled, Newton concentrated on mathematics, reading the works of Oughtred, Kepler, Viète, and Wallis. Within a few years he was recognized as the best mathematician in the world.

Legend has it that Newton was sitting under an apple tree eating his lunch when an apple fell on his head. Supposedly, at that very instant Newton formulated his *law of universal gravitation* which essentially states that any two bodies are attracted to one another. A friend of Newton later wrote that Newton himself repeated this story, but the legend, true or not, is incidental to the law.

With his law of gravitation Newton unified the study of motion of heavenly bodies and earthly bodies. Recall that the stars were previously thought of as quite different from earthly bodies; heavenly motion was considered to be eternal and perfect while earthly motion was considered quite variable and imperfect. The telescope had been in use since 1609, however, and the moon and planets were shown to have properties in common with the earth. Newton demonstrated that one of these properties was the theory of gravitation which applied to both heavenly and earthly motion. He said that *any* two bodies attract one another with a force ($F$) given by the formula

$$F = G \frac{m_1 m_2}{r^2}$$

where $m_1$ and $m_2$ are the masses of the bodies, $r$ is the distance between them, and $G$ is a constant. The fact that Newton's calculations of the motions of the moon and of objects at the surface of the earth agreed with experimental results was instrumental in convincing him that his theory was valid. In these calculations the attraction of the sun on the moon, of secondary importance to that of the earth on the moon, was neglected.

Newton published his law of universal gravitation in 1687 in his great work, *Mathematical Principles of Natural Philosophy*, though he had discovered it over 20 years earlier. In this book, Newton also gave his three famous *laws of motion* which made mechanics a mathematical science and have influenced scientific thought ever since. Mathematicians of the 1700s noted sadly that there was only one system of the world and since Newton had discovered it, there was nothing left for them to do; such great progress had been made in science that continual progress came to be expected. Yet, in the twentieth century it has been shown that there is no one system of the world. Even Newton's laws cannot account for all phenomena.

Newton's second law is a particularly important one for mathematics. It states that the change in motion of a body is proportional to the motive force impressed. This law is commonly expressed by the formula

$$F = ma$$

where $F$ is the force, $m$ is the mass of the body, and $a$ is the acceleration of the body. The acceleration is the rate of change in the velocity. It takes no force to maintain a body at constant velocity, but it does take a force to change its velocity. It is known from calculus that acceleration is the second derivative of position. Newton's second law can, therefore, be written as a differential equation

$$F = m\frac{d^2r}{dt^2}$$

where $r$ gives the position of the body. Thus, Newton's law has provided many mathematical problems for countless mathematicians who have applied it to many specific types of motion.

Newton made great contributions to calculus. He invented what he called his method of *fluxions* from the idea of *flowing*, or variable, quantities and their rates of *flow*. Newton thought of a curve as generated by a moving point, and he talked about the rate of change of the $y$ magnitude as the $x$ magnitude changed. He gave general rules for finding derivatives, much like we use now and much like Fermat used before him.

Consider the curve $y = x^2$ (fig. 5.10) with two neighboring points, $(x,y)$ and $(x + h, y_1)$. Since $y = x^2$, the coordinates of these points can also be represented as $(x, x^2)$ and $(x + h, (x + h)^2)$, respectively. The ratio giving the change in $y$ to the change in $x$ from one point to the next is, therefore,

$$\frac{(x+h)^2 - x^2}{x+h-x} = \frac{x^2 + 2hx + h^2 - x^2}{h} = \frac{2hx + h^2}{h}$$
$$= 2x + h$$

Seventeenth Century Origin

Newton wrote that the ultimate ratio, obtained when $h$ is put equal to zero, gives the instantaneous rate of change at the point $x$. Thus, he let $h$ be zero and concluded that the fluxion of $y = x^2$ is $\dot{y} = 2x$. Newton used the dot to indicate the fluxion, so he wrote $\dot{y}$ for the fluxion, rate of change or derivative, of $y$.

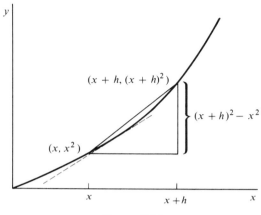

Figure 5.10

In order to find the rate of change of $y$ at the point $x$, Newton first chose another point $x + h$ and computed the average rate of change between $x$ and $x + h$ from the ratio of the change in $y$ to the change in $x$ over that interval. Then as the point $x + h$ moved closer to $x$, the ratio approached more closely the rate of change at $x$. Finally, when $h$ became zero, the rate of change at the point $x$ itself was obtained. It certainly seems reasonable that if we compute the average rate of change of $y$ over increasingly smaller intervals, it should approach the rate of change of $y$ at the point $x$ itself.

This process of finding the rate of change also has a geometric interpretation. The ratio giving the rate of change between $x$ and $x + h$ represents the slope of the line through $(x, x^2)$ and $(x + h, (x + h)^2)$, and as the point $x + h$ moves toward $x$, that line (slope) moves toward the tangent line. Thus, the rate of change of $y$ with respect to $x$ is given by the slope of the tangent to the graph at $x$. If $y$ is changing rapidly, then the slope is steep.

Newton's method was quite successful and was applied to solve many important problems, yet his justification of the process was not clear to many people. If one tries to interpret this calculation by the rules of algebra, the calculation does not make sense. **Bishop George Berkeley (1685–1753)** made just such a criticism. He was upset when mathematicians criticized religious thinking, so he pointed out that mathematicians' reasoning was the most nonsensical of all. For example, consider again the equation

$$\frac{(x+h)^2 - x^2}{x+h-x} = 2x + h$$

Recall that Newton let $h = 0$ in the final expression, $2x + h$, to obtain the ultimate ratio. Berkeley reasoned that $h$ should not be set equal to zero on just the right side of the equation. As we know from algebra, both sides of an equation must be treated equally. Therefore, if $h = 0$ on the right side of an equation, $h$ must be equal to 0 on the left side, also. Then

$$\frac{x^2 - x^2}{x - x} = \frac{0}{0}$$

which is meaningless.

The answer to Bishop Berkeley's criticism is, very simply, that calculus is not algebra. It involves something more — the idea of one point approaching another. This idea of a limit was not clarified for over 100 years after the time of Newton. Newton's intuitive approach was quite satisfactory for the curves to which he applied his calculus, but the material in modern calculus books consists of much theory which was developed later in response to more complicated problems.

In the previous calculation $(x + h)^2$ had to be expanded, a relatively simple procedure. For example, in computing the derivative of the curve $y = x^n$ we must expand $(x + h)^n$. This is done using the binomial theorem,

$$(a + b)^n = a^n + \binom{n}{1}a^{n-1}b + \binom{n}{2}a^{n-2}b^2 + \cdots + b^n$$

Newton extended the use of the binomial theorem to cases where $n$ was fractional and negative and used it in his derivative and integral calculations.

Finding the area of a region had traditionally been a difficult problem. The only method known was to use the method of Archimedes and approximate the region by simpler regions such as triangles or rectangles. Newton's great contribution to the solution of area problems was to make use of the idea that the rate of change of the area under a graph is equal to its height (if only the right endpoint varies). As an illustration, consider the curve in figure 5.11.

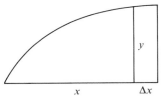

**Figure 5.11**

If we increase $x$ by $\Delta x$ the area increase is approximately $y \cdot \Delta x$, as the added region is almost rectangular in shape, particularly when $\Delta x$ is small. The rate of change in the area with respect to the change in $x$ is, thus, approximately $y \cdot \Delta x / \Delta x$ which equals the height $y$. The approximation becomes exact as the change, $\Delta x$, is reduced to zero. Newton expressed this result that the rate of change of area equals the height in formula as

$$\dot{A} = y$$

where $\dot{A}$ denotes the rate of change, or fluxion, of $A$. Using this formula, Newton could instantaneously solve what previously had been very difficult area problems. For example, given a curve $y = f(x)$ Newton had to think of a function having $y$ as its rate of change. Given $y = 2x$, Newton knew that $x^2$ has rate of change $2x$. Thus, the area function $A$ which satisfies $\dot{A} = 2x$ is given by $A = x^2$.

Newton simplified many other problems, also. He solved arc length problems by finding a formula $\sqrt{1 + \dot{y}^2}$ for the rate of change of arc length. Then the arc length itself was the function with that rate of change.

Newton made his methods much more powerful by combining them with his techniques of expanding functions into infinite series. Using infinite series he could solve problems that had been entirely impossible to solve before. He could divide to obtain

$$\frac{1}{1 + x^2} = 1 - x^2 + x^4 - x^6 + \cdots$$

and then find the area under the graph of $1/(1 + x^2)$, easily using the series. He could use the binomial theorem to find

$$\sqrt{1 - x^2} = 1 - \frac{1}{2}x^2 - \frac{1}{8}x^4 - \frac{1}{16}x^6 - \cdots$$

Newton's mathematical contributions were not confined to calculus. One of the results of Descartes' application of algebra to geometry was to simplify representation and discovery of new curves. Newton wrote a book on the graphs of cubic equations, showing the various forms these graphs might take. They are more interesting than the well-studied conics which are graphs of quadratic equations. Some examples are given in figure 5.12.

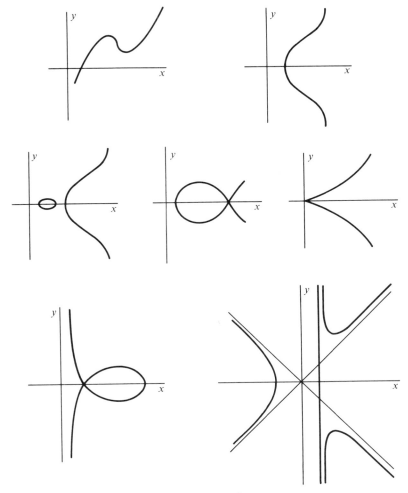

Figure 5.12

Newton was a great scientist and was fortunate enough to be recognized as an influential man by his contemporaries. He was interested in all aspects of science, particularly as they related to theology, and he tried to harmonize the dates of the Old Testament with those of secular history. Newton was also a member of Parliament under William and

Mary and Master of the Mint in 1699. In 1705 Newton was knighted by Queen Anne.

Great geniuses seem to have a reputation for complete concentration on a problem, and Newton is no exception. Many stories are told of his complete absorption in some deep problem. He was riding home from Grantham one day and dismounted to let his horse walk up a steep hill. At the top of the hill he turned to mount again, only to find that his horse had slipped away, leaving the empty bridle in Newton's hand. On another occasion, Dr. Stukeley, a friend of Newton, called on him. Newton was out, but the table was laid for dinner, so Dr. Stukeley took advantage of the situation and ate the meal. When Newton appeared later he greeted Stukeley and then prepared to eat. He lifted the covers of the dishes only to find that the food was gone. "Dear me," said Newton, "I thought I had not dined, but I see I have."[2] Perhaps stories like these are responsible for the term absentminded being applied to certain intellectuals.

Newton remained modest about his great contributions. He said, "If I have seen a little further than others it is because I have stood on the shoulders of giants." His laws had their roots in the studies of Galileo and Kepler. His powerful methods of calculus were made possible by the algebra which had been developing since the Middle Ages and its applications to geometry by Descartes, Fermat, and others prior to his time.

### GOTTFRIED WILHELM LEIBNIZ (1646-1716)

Gottfried Leibniz was a brilliant German who applied his genius to many fields — law, logic, philosophy, and mathematics. He was a lawyer by profession, and in his studies he became interested in natural philosophy. To understand natural philosophy he needed an understanding of math, so he studied mathematics with Huygens.

Leibniz made a project of reducing all exact reasoning to a symbolical technique. He wrote on the subject of symbolic logic, but its full development was not to take place until the nineteenth century. Leibniz also developed rules of differentiation and integration, noting that they are inverse processes. The notation which Leibniz used in his rules was very convenient — $dy/dx$ symbolized the derivative, and an elongated $S$ was the origin of the integral sign $\int$. The $S$ represented *sum*.

The notion of a function had been developing since the Middle Ages. In Descartes' geometry, curves were expressed by equations. Leibniz

---

[2]J. A. Holden, "Newton and his Homeland — The Haunts of his Youth," *Isaac Newton 1642-1727*, ed. W. J. Greenstreet (London: G. Bell and Sons, Ltd., for the Mathematical Association, 1927), pp. 142-43.

introduced the word *function* which referred to quantities such as tangents and normals that depend on a curve. A function was thought to be an algebraic relationship between the $x$ and $y$ variables, such as $x^2 + y^2 = 1$ or $y = 3x^2 + 2x$, or $y = 1 + x + x^2/2 + x^3/6 + \cdots$. The rules of calculus work quite well for such functions, so there was little need for intricate definitions of limit and continuity, except perhaps to clarify the concepts. Later, around 1800, this narrow notion of function was no longer adequate when applied to new problems, so the concept of function had to be extended.

Soon after the creation of the calculus, there arose disputes between supporters of Newton and supporters of Leibniz as to who first invented the rules of calculus. Some accused Leibniz of plagiarism, and the arguments were so fierce that mathematicians on the continent would not read English works and vice versa. This policy was detrimental to the English, because Leibniz's notation was easier to use than Newton's. Continental mathematics took the lead, therefore, in making new discoveries in calculus. The mathematical clash between the Germans and the English had a rather ironic twist, because Leibniz was employed by the man who became King George I of England. Queen Anne died childless in 1714, and the royal successor to the throne of England was a great-grandson of James I, George of Hanover.

## 2 Eighteenth Century Development

The late 1600s and the 1700s was the period when calculus began to flourish. Previous discoveries were combined with the inventions of many more basic techniques of calculus, and then this more sophisticated mathematics was applied to mechanics and other sciences. Newton's laws were applied to many physical problems and the mathematical problems arising from the applications were studied.

### MATHEMATICIANS OF THE EARLY 1700s

Among the mathematicians of the period are many names which the student of calculus will recognize. The following is a list of early eighteenth century mathematicians.

James (also Jakob or Jacques) Bernoulli   (1654–1705)
John (also Johann or Jean) Bernoulli   (1667–1748)
G.F.A. de L'Hospital   (1661–1704)

Abraham De Moivre   (1667–1754)
Colin Maclaurin   (1698–1746)
Gabriel Cramer   (1704–1752)
Michael Rolle   (1652–1719)
Jacopo Ricatti   (1676–1754)
Girolamo Saccheri   (1667–1733)
Brook Taylor   (1685–1731)

*L'Hospital's rule*, *Maclaurin's series*, *Cramer's rule*, *Rolle's theorem*, and *Taylor's series* are familiar terms to calculus students. Actually, only one of these five mathematicians was the original discoverer of the result attributed to him, and that man was Rolle. The person who popularizes a result generally has his name attached to it, although later it may be learned that someone else had originally discovered the same result. For practical purposes names are not changed, but, even so, the mistakes seem to compensate for one another. Although Maclaurin was credited with a series which he did not discover, a rule which he did originate is now known as Cramer's rule.

The Swiss Bernoulli brothers both contributed to the development of calculus. John posed, as a challenge to mathematicians, the problem of the *brachistochrone* (meaning *least time*). The problem was to find the curve between two given points, $A$ and $B$, along which a particle will slide in the least time (fig. 5.13). A cycloid was the desired curve. Both brothers solved the problem, but by different methods. The Bernoulli brothers were competitors in mathematics. John's son Daniel, also a mathematician, was viewed as a rival, too. John made Daniel move out of their house when he won a prize for which they were both competing.

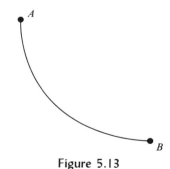

Figure 5.13

James Bernoulli wrote a book on probability, the *Art of Conjecture*, which contains some interesting results. Suppose a person invests $1 at 100% interest per year. At the end of the year he has $2 instead of $1. If interest is compounded every six months, then at six months he has $(1 + 1/2)$, and at one year he has $(1 + 1/2) + 1/2(1 + 1/2) =$

$(1 + 1/2)^2$. Continuing in this manner, it follows that if the interest is compounded $n$ times per year, then at the end of the year the total would be $(1 + 1/n)^n$. Does this amount increase without bounds as $n \to \infty$? Can the person become rich by having his interest compounded every second? Bernoulli showed that

$$2 < \left(1 + \frac{1}{n}\right)^n < 3$$

Thus, no matter how often interest is compounded on the $1, he will never receive more than $3. In fact, $(1 + 1/n)^n$ approaches the number $e = 2.71 \ldots$ as $n$ approaches infinity. Bernoulli proved the above inequality by expanding $(1 + 1/n)^n$ using the binomial theorem.[3]

John Bernoulli was acquainted with the Marquis de L'Hospital. In fact, L'Hospital was a rich man who agreed to support Bernoulli if Bernoulli would allow L'Hospital to publish Bernoulli's discoveries in calculus. Thus, the famous *L'Hospital's rule*, which appeared in the first textbook on calculus written by L'Hospital in 1696, was actually derived by Bernoulli. L'Hospital was also a mathematician, but he was not as capable as Bernoulli. After L'Hospital died, Bernoulli claimed his own discoveries, but there was no evidence to support him. It was not until 1955 that letters between Bernoulli and L'Hospital detailing the arrangement were found, thus proving that Bernoulli was correct in his claim.

## LEONHARD EULER (1707-1783)

Leonhard Euler of Switzerland was the most prolific writer of any mathematician; his collected works comprise almost 100 large volumes. He achieved this tremendous output, even though he was blind for the last 17 years of his life.[4]

In the 1700s the royal academies, rather than the universities, were the main centers of research in Europe. Two such institutions, both of which were founded by Leibniz, were the Berlin academy supported by Frederick the Great of Prussia and the St. Petersburg academy supported by Catherine the Great of Russia. Mathematicians and other specialists were paid to do research, as much for the prestige gained by the ruler as for practical applications to benefit the state. Euler went to St. Petersburg in 1727 to take a position on the medical research staff of the academy since there were no openings in mathematics. Soon after, however, he was able to transfer to the mathematics staff.

---

[3] See problem 21 of this chapter.
[4] See Bell, *Men of Mathematics*, pp. 139-152 for a biography of Euler.

**Eighteenth Century Development** 187

Euler had a great genius for manipulating formulas, as evidenced in his many books. He wrote on mechanics and the calculus of variations, the latter a subject in which he derived the fundamental equations. Euler's analytic geometry was almost identical to modern analytic geometry, and his calculus books were widely studied. Euler contributed much to the notation of calculus by introducing and/or popularizing the symbols, $e$, $\pi$, and $i$. He found a striking relationship,

$$e^{\pi i} = -1$$

which was based on the formula

$$e^{i\theta} = \cos\theta + i\sin\theta$$

In Euler's books the function concept came into its own. His definition of a function of a variable quantity (based on John Bernoulli's) is *any analytic expression whatsoever made up from that variable quantity and from numbers or constant quantities.* Thus,

$$y = x^7 + 4x^3 + 3$$

is a function, as is

$$y = 1 + x + \frac{x^2}{2} + \frac{x^3}{6} + \frac{x^4}{24} + \cdots$$

Much work in infinite series is also contained in Euler's writings. He found the sum of the reciprocals of the squares, a problem that Leibniz and James Bernoulli could not solve, to be

$$\frac{\pi^2}{6} = \frac{1}{1^2} + \frac{1}{2^2} + \frac{1}{3^2} + \frac{1}{4^2} + \cdots$$

Euler also was interested in number theory, and he proved the converse of Euclid's theorem about perfect numbers. Euclid had shown that if $2^n - 1$ is prime, then $2^{n-1}(2^n - 1)$ is perfect. Euler proved that all even perfect numbers are of the form $2^{n-1}(2^n - 1)$ where $2^n - 1$ is prime. It is not known to this day if there are any odd perfect numbers, or if there are infinitely many perfect numbers. In a letter to Euler, **Christian Goldbach (1690–1764)** suggested that every even integer greater than or equal to 4 is the sum of two primes, for example, $12 = 5 + 7$ and $32 = 19 + 13$. Again, this has been neither proved nor disproved. A similar problem concerns whether or not there are infinitely many pairs of twin primes (primes differing by two) such as 11 and 13, 17 and 19, or 29 and 31. This is also unsolved.

An interesting problem which Euler solved by what he called the *geometry of position* is the problem of the seven bridges of Königsberg.

The town of Königsberg had seven bridges because of the nature of the river which flowed through the town. The river forked, and there was an island in the river (fig. 5.14). Could someone follow a path so that he crosses each bridge once, but none more than once?

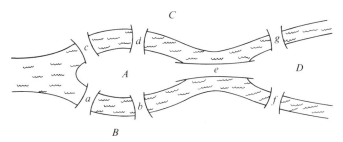

Figure 5.14

Euler proved that the answer is no. In fact, he solved the problem in the general case, showing that the key quantity is the number of land masses which have an odd number of bridges. If there are no land masses with an odd number of bridges, then such a crossing is possible starting anywhere. If there are two land masses with an odd number of bridges, however, a crossing is possible only by starting on one of these masses and ending on the other. If the number of land masses with an odd number of bridges is one or is greater than or equal to three, then the required crossing is impossible. In Konigsberg, the number of bridges at each mass is odd. Thus, crossing each bridge once and only once is impossible by Euler's criterion. However, if we remove bridge $e$, then only masses $B$ and $C$ have an odd number of bridges, and a crossing is possible starting on either mass $B$ or $C$ and ending on mass $C$ or $B$, respectively.

## JOSEPH-LOUIS LAGRANGE (1736-1813)

Lagrange, another outstanding mathematician of the 1700s, was a French-Italian. In 1770, Lagrange published an important paper on the solution of equations. Ever since 1545, formulas had been known for the general solution to third and fourth degree equations. Lagrange analyzed these solutions in the hope of finding general solutions to the quintic and higher degree equations. His methods were later used to prove that such general solutions are impossible. More importantly, his ideas led to the important concept of a group. Thus, his work is very valuable, in spite of the fact that he did not solve his original problem.

The general $n$th degree equation may be written as $x^n + a_1 x^{n-1} + a_2 x^{n-2} + \cdots + a_n = 0$ and has roots $x_1, x_2, \ldots, x_n$. For simplicity, consider a cubic $x^3 + ax^2 + bx + c = 0$ with roots $x_1, x_2,$ and $x_3$. Lagrange used the six permutations of these roots:

$$s_1: \quad x_1 \to x_1,\; x_2 \to x_2,\; x_3 \to x_3$$

$$s_2: \quad x_1 \to x_2,\; x_2 \to x_1,\; x_3 \to x_3$$

$$s_3: \quad x_1 \to x_3,\; x_2 \to x_1,\; x_3 \to x_2$$

$$s_4: \quad x_1 \to x_1,\; x_2 \to x_3,\; x_3 \to x_2$$

$$s_5: \quad x_1 \to x_3,\; x_2 \to x_2,\; x_3 \to x_1$$

$$s_6: \quad x_1 \to x_2,\; x_2 \to x_3,\; x_3 \to x_1$$

Notice that the permutation $s_1$, called the *identity permutation*, leaves each root unchanged, while $s_2$ interchanges $x_1$ and $x_2$, and so on.

Lagrange applied these permutations to functions of the roots. For example, consider some functions of the roots, such as

$$g_1(x_1, x_2, x_3) = x_1 + x_2 + x_3$$
$$g_2(x_1, x_2, x_3) = x_1 x_2 x_3$$
$$f_1(x_1, x_2, x_3) = x_1^2 + x_2 + x_3$$
$$f_2(x_1, x_2, x_3) = x_1 + x_2^2 + x_3$$
$$f_3(x_1, x_2, x_3) = x_1 + x_2 + x_3^2$$
$$h_1(x_1, x_2, x_3) = x_1 + \left(\frac{-1 + \sqrt{-3}}{2}\right)x_2 + \left(\frac{-1 - \sqrt{-3}}{2}\right)x_3$$
$$h_2(x_1, x_2, x_3) = x_1 + \left(\frac{-1 - \sqrt{-3}}{2}\right)x_2 + \left(\frac{-1 + \sqrt{-3}}{2}\right)x_3$$
$$d(x_1, x_2, x_3) = (x_1 - x_2)(x_1 - x_3)(x_2 - x_3)$$

If we apply any permutation to $g_1(x_1, x_2, x_3)$, the function remains unchanged. For example, applying $s_2$, we interchange $x_1$ and $x_2$ giving

$$g_1(x_2, x_1, x_3) = x_2 + x_1 + x_3$$

which equals $x_1 + x_2 + x_3$. You can check that all permutations leave $g_2$ unchanged, also, but consider the effect of each of the six permutations on $f_1$.

$$s_1 f_1 = f_1(x_1,x_2,x_3) = x_1^2 + x_2 + x_3 = f_1$$
$$s_2 f_1 = f_1(x_2,x_1,x_3) = x_2^2 + x_1 + x_3 = f_2$$
$$s_3 f_1 = f_1(x_3,x_1,x_2) = x_3^2 + x_1 + x_2 = f_3$$
$$s_4 f_1 = f_1(x_1,x_3,x_2) = x_1^2 + x_3 + x_2 = f_1$$
$$s_5 f_1 = f_1(x_3,x_2,x_1) = x_3^2 + x_2 + x_1 = f_3$$
$$s_6 f_1 = f_1(x_2,x_3,x_1) = x_2^2 + x_3 + x_1 = f_2$$

Two permutations, $s_1$ and $s_4$, leave $f_1$ unchanged. Lagrange called $\{s_1, s_4\}$ the *group* of the function $f_1$. This is the original meaning of the word group; the group of a function was that set of permutations of the roots which left the function unchanged. Thus, $\{s_1, s_2, s_3, s_4, s_5, s_6\}$ is the group of $g_1$. The functions $f_2$ and $f_3$ are called *conjugates* of $f_1$ — the two permutations $s_2$ and $s_6$ change $f_1$ into its conjugate $f_2$, while $s_3$ and $s_5$ change $f_1$ into $f_3$. You might check to see that the group of $h_1$ is $\{s_1\}$, while the group of $d$ is $\{s_1, s_3, s_6\}$.

Some functions such as $g_1$ are left invariant under every permutation while others such as $h_1$ are left invariant only by the identity permutation, $s_1$. Functions such as $g_1$ are called *symmetric*. Given an equation, symmetric functions of the roots are easy to find. To find $g_1(x_1,x_2,x_3)$, first write the general equation of a cubic, $x^3 + ax^2 + bx + c = 0$, as a product of linear factors,

$$x^3 + ax^2 + bx + c = (x - x_1)(x - x_2)(x - x_3)$$

Expanding the right-hand side of the equation gives

$$x^3 + ax^2 + bx + c$$
$$= x^3 - (x_1 + x_2 + x_3)x^2 + (x_1 x_2 + x_2 x_3 + x_1 x_3)x - x_1 x_2 x_3$$
$$= 0$$

In order for the left side to agree with the right side, the following relationship must exist.

$$g_1(x_1,x_2,x_3) = (x_1 + x_2 + x_3) = -a$$

Thus, $g_1(x_1,x_2,x_3)$ is known immediately at the beginning of the problem, because $(x_1 + x_2 + x_3)$ is the negative of the coefficient of the $x^2$-term.

The function $h_1$, as noted above, is nonsymmetrical and is difficult to find, in contrast to $g_1$ which is easy to find. Yet, knowing the values $h_1(x_1,x_2,x_3)$ and $h_2(x_1,x_2,x_3)$, we could then find the roots $x_1$, $x_2$, $x_3$ of the cubic equation. Suppose that $h_1(x_1,x_2,x_3) = B$ and $h_2(x_1,x_2,x_3) = C$. Using the definitions of $g_1$, $h_1$, and $h_2$ we obtain the equations

$$x_1 + x_2 + x_3 = -a$$

$$x_1 + \left(\frac{-1 + \sqrt{-3}}{2}\right)x_2 + \left(\frac{-1 - \sqrt{-3}}{2}\right)x_3 = B$$

$$x_1 + \left(\frac{-1 - \sqrt{-3}}{2}\right)x_2 + \left(\frac{-1 + \sqrt{-3}}{2}\right)x_3 = C$$

which can be solved for the three unknowns, $x_1$, $x_2$, and $x_3$. Although solving these equations is cumbersome, the end results are general formulas for the three roots of the cubic equation, analogous to the quadratic formula, $x = (-b \pm \sqrt{b^2 - 4ac})/2a$, for the two roots of the equation $ax^2 + bx + c = 0$. Thus, the cubic equation can be solved provided the values of $h_1$ and $h_2$ can be determined.

Lagrange's method for finding functions such as $h_1$ was based on the following theorem which he proved.

**Theorem**   Let $z$ and $w$ be any two functions of the roots, and let $G$ be the group of $z$, namely, the set of all permutations leaving $z$ invariant. Suppose that the function $w$ is not left invariant by all the permutations of $G$, but rather has $r$ conjugates, including itself. If the value of $z$ is known, then the value of $w$ can be determined by solving an equation of degree $r$.

One could apply this theorem to the functions $g_1$ and $h_1$. The function $g_1$ is invariant under all six permutations of the roots, while the function $h_1$ has six different conjugates, including itself, under the application of these same six permutations. Thus, the theorem states that $h_1$ can be determined by solving an equation of degree six, but such an equation represents a more difficult problem than the original which was to solve a cubic equation.

Lagrange found it helpful to add an intermediate step. He first applied his theorem to the functions $g_1$ and $d$. The function $d$ has two values when the six permutations leaving $g_1$ invariant are applied to it, and, thus, can be found by solving a quadratic equation. Once $d$ was known, Lagrange could then find $h_1$. The three permutations leaving $d$ invariant transform $h_1$ into three different functions, so that, according to the theorem, $h_1$ can be found by solving a cubic equation, which in this case happens to be of a simple form that can be solved easily by taking a cube root. Thus, by inserting another step, Lagrange was able to find the function $h_1$, and in the same manner $h_2$, and so solve the cubic. What he had done was to thoroughly analyze Cardano's solution to the cubic, in which the first step was to solve a quadratic equation and the second was to extract a cube root.

Lagrange's goal was to extend his use of his theorem so that he could solve fifth and higher degree equations. In order to solve the fifth degree equation, Lagrange looked for a sequence of functions, like $g_1 \to d \to h_1$ in the cubic case, so that he could reduce the solution to a sequence of easy steps. Unfortunately, he could not find such a sequence of functions for the fifth degree equation. It was later proved, using Lagrange's ideas, that a solution by roots to the fifth degree equation is, in general, impossible.

The theory of equations is a topic of pure mathematics, yet Lagrange also made significant contributions to applied mathematics. Typically, the most brilliant mathematicians have excelled in several areas of mathematics. Lagrange, at age 19, wrote a book *Analytical Mechanics* which he was proud to say contained no diagrams. This work was not published until 1788 when he was 52. Lagrange derived some basic results in the calculus of variations which he used to obtain his equations of mechanics. You may have heard his name in connection with the *Lagrange multipliers* used in solving certain maximum and minimum problems. Lagrange became the first mathematics professor in the new École Polytechnique, founded in 1797 to train engineers for Napoleon.

## PIERRE-SIMON DE LAPLACE (1749-1827)
## AND GASPARD MONGE (1746-1818)

**Laplace,** a Frenchman, concentrated on the application of the law of gravitation to the entire solar system. He wanted to determine if the solar system was stable, and he showed that it was, if it conformed to his mathematical model. Laplace's major work, *Celestial Mechanics*, contains his ideas concerning the solar system. Laplace was an applied mathematician — he was primarily concerned with explaining phenomena and not with mathematical theory itself. A pure mathematician is more interested in the mathematical theory.

Laplace's name is associated with several important topics in mathematical physics. The *Laplacian equation* is perhaps the most important partial differential equation

$$\frac{\partial^2 u(x,y,z)}{\partial x^2} + \frac{\partial^2 u(x,y,z)}{\partial y^2} + \frac{\partial^2 u(x,y,z)}{\partial z^2} = 0$$

This equation describes gravitational potential and many other quantities. Many problems of mathematical physics can be formulated as partial differential equations.

**Monge,** also a Frenchman, invented descriptive geometry, a method of representing a three-dimensional object on a two-dimensional surface. While in engineering school Monge used descriptive geometry to solve a problem of a fort design. He arrived at the solution so quickly that the teacher did not believe that he had found the answer. When the teacher was persuaded to check the solution, however, he found Monge was correct. Consequently, Monge was made a teacher of descriptive geometry. He was sworn not to divulge his method in order that he alone might teach it, so for fifteen years descriptive geometry was a military secret.[5] This subject now is a part of drafting.

Monge was a relatively important official in the French Revolutionary government under Napoleon. When Napoleon journeyed to Egypt, Monge accompanied him on the expedition as a member of a Legion of Culture which was to improve the minds of the Egyptians. The mission did not fare too well, however, so the men returned to France.

Western mathematics and science were by 1800 well along on the journey begun almost two hundred years previously with the development of analytic geometry and, shortly after, of the calculus. Just as the knowledge from physical explorations altered people's conceptions of their world, so did these mental explorations change people's ideas of the universe and of their place in it.

---

[5]Ibid., p. 185.

## Problems

1. Descartes showed how to perform arithmetic operations geometrically. Suppose the product of *BC* and *BD* is required (fig. 5.15). Let *AB* be the unit length segment. Connect *AC*. Construct *DE* parallel to *AC*.

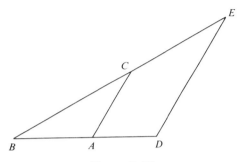

Figure 5.15

   a. Show that the length of *BE* is the product of the lengths of *BC* and *BD*.
   b. Use the above method to construct the product of 2 and 3. Check to see that it is approximately 6.

2. Descartes also performed division geometrically.
   a. Referring to figure 5.15, show that if *BE* and *BD* are given, if $AB = 1$, and if *AC* is constructed parallel to *DE*, then $BC = BE/BD$.
   b. Use the method in part a to construct the quotient of 8 divided by 4.

3. Descartes constructed the positive solutions to quadratic equations. Given the equation $z^2 = az + b^2$ he constructed a right triangle *NLM*, with *LM* equal to $b$, the square root of the constant term, and *LN* equal to $1/2a$, half the coefficient of $z$. Construct the circle with center *N* and radius *NL*, and extend *MN* to *O* (fig. 5.16).
   a. Show that $z = OM$ is the positive solution to the equation. [*Hint:* Use the quadratic formula to solve for $z$. Compare the result with *OM*.]
   b. Construct the positive solution to $z^2 = 4z + 9$.

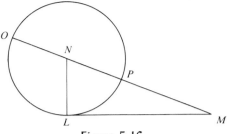

Figure 5.16

4. a. Given the equation $y^2 = -ay + b^2$ and figure 5.16, show that $y = MP$ is the positive solution to the equation.
   b. Construct the positive solution to $y^2 = -3y + 16$.

5. To solve the equation $z^2 = az - b^2$, Descartes constructed a right angle $NLM$ with $NL$ equal to $1/2a$ and $LM$ equal to $b$. He then constructed a circle with center $N$ and radius $NL$ and constructed a line through $M$ parallel to $NL$. (See fig. 5.17.)

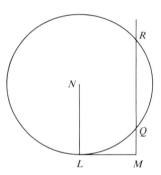

Figure 5.17

a. Show that the lengths $MQ$ and $MR$ represent the two positive solutions to the given equation. [*Note:* Here $NL$ is assumed greater than $LM$, so that the circle and line do, in fact, intersect.]

b. Construct the two solutions to the equation $z^2 = 6z - 4$.

c. If $NL$ is less than $LM$, then the circle and the line do not intersect, and no solutions are obtained by this method. Explain this phenomenon. [*Hint:* You may wish to refer to a specific equation, such as $z^2 = z - 4$.]

6. As an example of a three-line locus problem, let the lines $x = 1$, $x = -1$, and $y = 0$ be given (fig. 5.18). Find the equation of and sketch the locus of points $P = (x, y)$, such that $PQ \cdot PR = PS$.

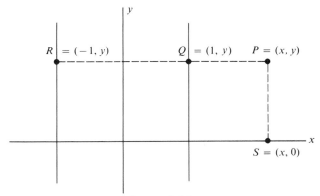

**Figure 5.18**

7. Let the four lines $x = 1$, $x = -1$, $y = 1$, $y = -1$ be given (fig. 5.19). Find the equation of and sketch the locus of points $P = (x,y)$, such that $PQ \cdot PR = PS \cdot PT$.

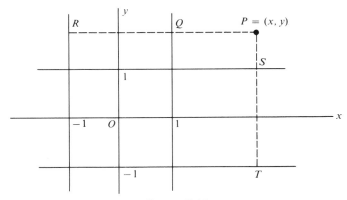

**Figure 5.19**

8. Let the four lines $x = 0$, $x = 1$, $x = -1$, and $y = 0$ be given (fig. 5.20). Find the equation of and sketch the locus of points $P = (x,y)$, such that $PQ \cdot PS = PR \cdot PT$.

9. Descartes used a rule of signs, which has since been named after him, to find bounds for the numbers of positive and negative roots of an equation. The rule states that an equation can have as many positive roots as it contains changes of sign from positive to negative or from negative to positive and as many negative roots as the number of times that two positive signs or two negative signs are found in succession. For example, the equation

$$x^5 - 3x^4 + 2x^3 + 3x^2 - x - 1 = 0$$

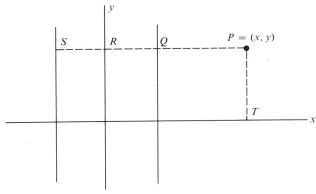

Figure 5.20

can have, at most, 3 positive roots and 2 negative roots, as its sequence of signs is $+, -, +, +, -, -$. Find the maximum numbers of positive and negative roots for the following equations.
  a. $3x^2 - 2x - 1 = 0$
  b. $x^4 + x^3 + x^2 - 2x + 1 = 0$
  c. $-x^5 - 3x^4 + x^3 - x^2 + 3x + 5 = 0$

10. According to Fermat's theorem, what is the smallest prime which could possibly divide $2^{11} - 1$? Does it?

11. a. Use Fermat's theorem to show that if $2^{13} - 1$ is not divisible by 53 or 79, then it is prime. [*Hint:* Recall that if $N = ab$, either $a \leq \sqrt{N}$ or $b \leq \sqrt{N}$, so that if a number $N$ has no divisors less than or equal to $\sqrt{N}$, then $N$ is prime.]
    b. Show that $2^{13} - 1$ is prime, and that $2^{12}(2^{13} - 1)$ is, therefore, a perfect number.

12. Let an odd prime $p$ be the sum of two squares, $p = s^2 + t^2$.
    a. Show that both $s$ and $t$ cannot be even.
    b. Show that both $s$ and $t$ cannot be odd.
    c. From parts a and b, either $s$ or $t$ is even. Suppose $s$ is even and $t$ is odd. Show that $p$ is always one greater than a multiple of four. Thus, any prime $p$ which is one less than a multiple of four cannot be written as a sum of two squares. [*Hint:* Let $s = 2m$ and $t = 2n + 1$ and show that $p$ is one greater than a multiple of four.]

13. Prove Desargues' theorem for the case in which the triangles are not located in the same plane. [*Hint:* Consider the intersection of the planes containing each triangle.]

14. Show that the cross-ratio of the points $A$, $B$, $C$, and $D$ is equal to the cross-ratio of the points $A'$, $B'$, $C'$, and $D'$ in figure 5.21.

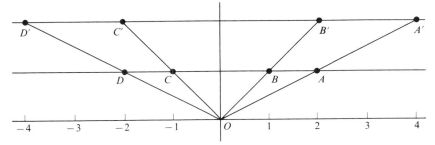

Figure 5.21

15. Roberval found the area under an arch of the cycloid in an interesting way, as illustrated in figure 5.22. Let $ECI$ be half an arch of a cycloid. Roberval defined a curve $EDHI$ called the companion to the cycloid on which a point $D$ has the property that $CD = AB$.

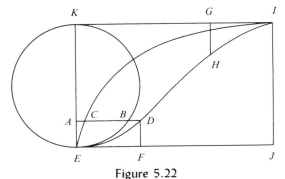

Figure 5.22

   a. Show that the area between the cycloid half and its companion is equal to the area of the semicircle.
   b. It can be shown that the companion bisects the rectangle $EJIK$. Show that the area of the rectangle $EJIK$ is twice the area of the circle.
   c. Show that the area under the cycloid half is 3/2 the area of the circle, and, thus, that the area of the cycloid is three times the area of the circle.

16. Suppose $A$ and $B$ each bet $1 on the outcome of five coin tosses. Gambler $A$ bet that heads would occur three or more times, while $B$ bet that tails would. After two tosses, both resulting in heads, the game was stopped. How should the stakes have been divided between $A$ and $B$, so that each would receive his fair share?

17. Use the formula arctan $x$ + arctan $y$ = arctan $(x + y)/(1 - xy)$ to show that
   a. 2 arctan $1/5$ = arctan $5/12$
   b. 3 arctan $1/5$ = arctan $37/55$
   c. 4 arctan $1/5$ = arctan $120/119$
   d. 4 arctan $1/5$ = arctan $1/239$ + arctan 1, or
      $\pi/4$ = 4 arctan $1/5$ − arctan $1/239$

18. Use the formula in problem 17d to compute an approximation to $\pi$. [*Hint:* To compute arctan $1/5$, use the first three terms of its infinite series, and use the first term of the series for arctan $1/239$ to compute it.] Compare your answer with the value $\pi$ = 3.141592 correct to six places.

19. Divide 1 by $1 + x^2$ using ordinary polynomial long division to obtain an infinite series equal to $1/(1 + x^2)$.

20. Using the same method, divide 1 by $1 - x$ to obtain an infinite series equal to $1/(1 - x)$.

21. Show that $2 < [1 + (1/n)]^n < 3$. [*Hint:* Using the binomial theorem, show that

$$\left(1 + \frac{1}{n}\right)^n < 1 + 1 + \frac{1}{2} + \frac{1}{2 \cdot 3} + \frac{1}{2 \cdot 3 \cdots n}$$

$$< 1 + 1 + \frac{1}{2} + \frac{1}{2^2} + \cdots + \frac{1}{2^n}\Big]$$

22. Refer to figure 5.23 and determine if a person can walk across each bridge once and only once.

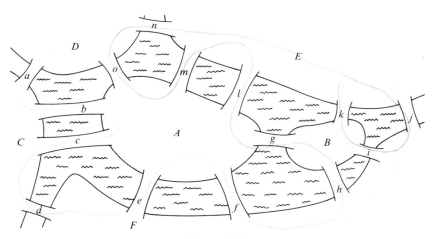

Figure 5.23

23. In the Königsberg bridge problem (fig. 5.14), a walk can be described by specifying the land masses traversed in order. Thus, $ACDA$ represents a walk from mass $A$ to $C$ to $D$ and back to $A$ in which three bridges are crossed. Suppose all seven bridges could be crossed exactly once.
   a. How many letters long must a description of such a walk be?
   b. Five bridges go to land mass $A$. If each bridge is crossed exactly once during the walk, how many times must the letter $A$ occur in the description of the walk?
   c. Three bridges go to each of masses $B$, $C$, and $D$. If each of these bridges is crossed exactly once during the walk, how many times must each of the letters $B$, $C$, and $D$ occur in the description of the walk?
   d. Comparing the number of occurrences of the letters $A$, $B$, $C$, and $D$ required in parts b and c with the total number allowed in part a, what can you conclude about the possibility of such a walk?

24. Suppose that $N$ bridges connect a certain group of land masses, and that each mass has an even number of bridges. Suppose trips are described, as in problem 23, by giving the sequence of land masses traversed. Let each bridge be crossed exactly once.
   a. If the trip does not start on a certain land mass $X$, how many times must the letter $X$ occur in the description of the walk?
   b. If the trip does start at a certain land mass $Y$, how many times must the letter $Y$ occur in the description of the walk?
   c. Use parts b and c to show that the number of letters in the description of the trip is $N + 1$, exactly the number required for any trip crossing each of $N$ bridges exactly once. Thus, such a trip is possible. (Recall that since each bridge has two ends, $N$ bridges have $2N$ ends.)

25. The equation $x^2 + 2x - 3 = 0$ with roots $-3$ and $1$ can be factored as $(x + 3)(x - 1) = 0$. Euler assumed this property held for polynomials of infinite degree to prove his formula

$$\frac{\pi^2}{6} = \frac{1}{1^2} + \frac{1}{2^2} + \frac{1}{3^2} + \cdots$$

Euler knew that

$$\sin z = z - \frac{z^3}{3!} + \frac{z^5}{5!} - \frac{z^7}{7!} + \cdots$$

and that the roots of $\sin z = 0$ are $z = 0, \pm\pi, \pm 2\pi, \pm 3\pi, \pm 4\pi, \ldots$.

a. Show that all roots of $\sin z = 0$, except $z = 0$, satisfy the equation

$$0 = 1 - \frac{z^2}{3!} + \frac{z^4}{5!} - \frac{z^6}{7!} + \cdots$$

b. Assuming that $\sin z$ with its infinite number of roots can be factored like a polynomial, show that

$$0 = \left(1 - \frac{z^2}{\pi^2}\right)\left(1 - \frac{z^2}{4\pi^2}\right)\left(1 - \frac{z^2}{9\pi^2}\right)\cdots$$

c. Assuming that the infinite product in part b can be multiplied out in the same way as a finite product, find the coefficient of $z^2$ in that product and show that by setting it equal to the coefficient of $z^2$ in the sum in part a, Euler's formula is obtained.

26. Verify Goldbach's conjecture for every even integer between 4 and 50.

27. For each of the following functions, find the group of permutations which leave it invariant.
 a. $f(x_1, x_2, x_3) = x_1 x_2 + x_3$
 b. $d(x_1, x_2, x_3) = (x_1 - x_2)(x_1 - x_3)(x_2 - x_3)$
 c. $h_1(x_1, x_2, x_3) = x_1 + [(-1 + \sqrt{-3})/2]x_2 + [(-1 - \sqrt{-3})/2]x_3$

28. Show that if $n$ is composite ($n = rs$, where $r > 1$ and $s > 1$), then $2^n - 1$ is composite. [*Hint:* Factor $(2^r)^s - 1$.]

29. Use problem 28 and Euler's theorem about perfect numbers to show that if $n$ is composite, then $2^{n-1}(2^n - 1)$ is not perfect.

30. Let $N$ be an even perfect number, so that $N = 2^{n-1}b$. Let $S$ be the sum of the positive divisors of $b$.
 a. Show that $N = (1 + 2 + 2^2 + \cdots + 2^{n-1})S - N$.
 b. Show that $2^n b = (2^n - 1)S$. [*Hint:* For any $n$, $1 + 2 + 2^2 + \cdots + 2^{n-1} = (2^n - 1)/(2 - 1)$]
 c. Show that $S = b + b/(2^n - 1)$ and, therefore, that $b/(2^n - 1)$ is an integer and a divisor of $b$.
 d. Show that $b = 2^n - 1$. [*Hint:* The number 1 is a divisor of $b$.]
 e. Show that $b$ is prime. Thus, Euler's theorem which states that if $N$ is an even perfect number, then $N = 2^{n-1}(2^n - 1)$ where $2^n - 1$ is prime, is proved.

# References

Seventeenth Century Origins

    Beckmann                        Newton
    Bell                              Smith (2)
    Boyer (1)                       Struik (3)
    Descartes                     Whiteside
    More

Eighteenth Century Development

    Bell                              Smith (2)
    Boyer (1)                       Struik (3)
    Dehn                          Wolff
    Dörrie

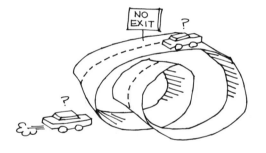

# 6 Mathematics as Free Creation

Men of the 1600s, such as Kepler and Galileo, believed that mathematics was the key to truth; nature was assumed to operate according to mathematical laws. Since the early 1800s, however, the spirit of modern thought has prevailed, the continuing realization that mathematical theories are independent of physical phenomena. Just because all physical triangles seem to have 180° as the sum of their angles, does not mean that this is true of every triangle in geometry. If geometry is analogous to the real world, then we say it provides a model of the world, just as a wooden ship in someone's den provides a model of an actual ship on the ocean. Suppose one decides, in a moment of inspiration, to build a model of a ship which does not exist, in fact, a model so different that the corresponding "real" ship would not float. A critic's first reaction might be that this is a waste of time since the model does not correspond to reality. Some genius may discover that the "unreal" ship, while not feasible as a ship, does, in fact, provide an excellent undersea habitation. On the other hand, he may discover an ocean, perhaps on a distant planet, on which the "unreal" ship would be quite suitable as a floating vessel.

When mathematics became independent of the physical world, it became free. This freedom allowed mathematics to develop in unforeseen ways. New theories were created with new models and many new applications. No longer could one limit the definition of mathematics to the study of number and magnitude. We shall see this freedom in geometry and algebra. In calculus also, intuitive notions were found to be unreliable, and a more formal logical basis for calculus was sought.

# 1  A Forerunner — Carl Friedrich Gauss (1777-1855)

Carl Friedrich Gauss, a German, was one of the greatest mathematicians of all time and certainly the greatest of his time. He anticipated many of the advances of the nineteenth century.

Gauss was an extremely bright child. He taught himself to read and to make mathematical calculations, and before the age of three, he corrected his father's addition of the payroll for his workers. When Gauss was 10 years old, he developed the formula $n(n + 1)/2$ for the sum of the first $n$ integers. Gauss' elementary school teacher asked the students in Gauss' class to find the sum of the numbers from 1 to 100, expecting to keep them busy for a long time. But Gauss realized that the numbers could be grouped as $1 + 100$, $2 + 99$, $3 + 98$, and so on up to $50 + 51$. Each of these sums equals 101, and there are 50 sums, so the sum of the first 100 integers is equal to $101 \times 50$, a simple product which Gauss calculated immediately. The assistant teacher recognized Gauss' extraordinary ability and gave him extra help and encouragement. He persuaded the wealthiest man in town, the Duke of Brunswick, to support Gauss' education.

At age 18, Gauss invented the *method of least squares* for finding the best value of a sequence of measurements of the same quantity. Another use of the method of least squares involves finding the best straight line fit to a set of data. To find the best line, one minimizes the sum of the squares of the differences of the data from the line (fig. 6.1). If we just added the differences above and below the line, there would be cancellation of the plus and minus terms. When the differences are squared, all terms are positive, so no cancellation can occur.

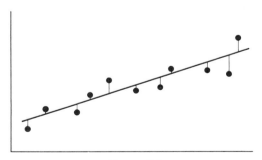

Figure 6.1

Gauss later used his method of least squares in fitting orbits of planets to data measurements, a procedure which was published in 1809 in his *Theory of motion of the heavenly bodies*. In connection with the error of observations, he developed the standard normal curve. Gauss derived the normal curve by assuming that if a sequence of measurements of the same quantity are given, the most probable correct value of that quantity is the average of all the measurements. Thus, figure 6.2 shows that the probability is greatest for the average of all the measurements and decreases for measurements either larger or smaller than the average, since such measurements, as they get further from the average, are less likely to be the true value of the measured quantity. The probability distribution which the curve represents is called the *Gaussian distribution*, the most useful distribution in probability theory and statistics.

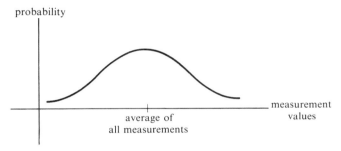

**Figure 6.2**

Just before 1800, complex numbers were given a geometric interpretation which made mathematicians feel more comfortable using them. Gauss graphed complex numbers in the plane, as did **Jean Argand (1769–1822)**, a Frenchman, and **Caspar Wessel (1745–1818)**, a Norwegian surveyor. Wessel was the first to publish his discovery in 1797. He represented a complex number such as $2 + 3\sqrt{-1}$ by an arrow from the origin to a point 2 units in the positive *x*-direction (real axis) and up 3 units in the *y*-direction (imaginary axis) (fig. 6.3). Each complex number represented a rotation and a "stretching." For example, the number $1 + \sqrt{-1}$

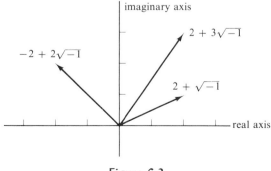

**Figure 6.3**

is $\sqrt{2}$ units long at a 45° angle with the positive real axis. When another complex number is multiplied by $1 + \sqrt{-1}$, it is rotated by 45° and multiplied in length by $\sqrt{2}$. Let $\sqrt{-1}$ be multiplied by $1 + \sqrt{-1}$. The number $\sqrt{-1}$ has length one and is at a 90° angle with the positive real axis. The product $(1 + \sqrt{-1})\sqrt{-1}$ should, therefore, have a length of $\sqrt{2}(1) = \sqrt{2}$ and be at an angle of $90° + 45° = 135°$. The product is $-1 + \sqrt{-1}$, and it is of length $\sqrt{2}$ at an angle of 135° (fig. 6.4). Thus, complex numbers were understood by giving them a new interpretation, just as negatives were first understood as a deficit or loss. A negative number of apples does not make any sense. Neither does a square whose area is negative. But given a new interpretation, both negatives and square roots of negatives seem reasonable.

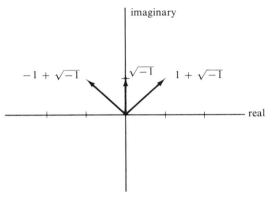

Figure 6.4

Gauss used the geometrical representation of complex numbers in proving a surprising result about regular polygons. The triangle and the pentagon were the only regular polygons with a prime number of sides which Euclid knew to be constructible with a straightedge and compass. No one since had added any others. Gauss proved that if $p = 2^{2^n} + 1$ is prime, then the $p$-sided polygon is constructible. Since $17 = 2^{2^2} + 1$ is prime, the 17-sided polygon is constructible using a straightedge and compass. The only known primes of the form $2^{2^n} + 1$ are 3, 5, 17, 257 and 65, 537. The 257-sided polygon has been constructed, and someone spent years in trying to construct the 65,537-sided polygon, but he did not succeed. Many of the numbers $2^{2^n} + 1$ have been found to be composite, for example, in 1958, $2^{2^{1945}} + 1$ was found to be divisible by $5 \times 2^{1947} + 1$. If written out in numerals of a length of one centimeter each, this number would be many times the circumference of the known universe.

Gauss used algebra to find the new constructible regular polygons. Since the square root of any quantity can be constructed with a straightedge and compass, any expression containing only square roots and numbers can be constructed with a straightedge and compass. For example, to construct $\sqrt{34 + 2\sqrt{17}}$, start with a line of length 17, then construct the line $\sqrt{17}$, then the line $2\sqrt{17}$, then the line $34 + 2\sqrt{17}$, then the line $\sqrt{34 + 2\sqrt{17}}$. (See problems 1 and 2 of chapter 5 and problem 3 of this chapter for some construction methods.)

Consider the unit circle with an inscribed 17-sided polygon. The angle $\theta$ (fig. 6.5) subtended by its side is $(360/17)°$. The cosine of $(360/17)°$ is the line $OT$ in figure 6.5. If $OT$ is constructible, then by constructing a perpendicular at the point $T$ the point $P$ can be found, and the side of the 17-sided regular polygon can be obtained by joining $PS$. Gauss was able to show that

$$\cos\left(\frac{360}{17}\right)° = -\frac{1}{16} + \frac{1}{16}\sqrt{17} + \frac{1}{16}\sqrt{34 - 2\sqrt{17}}$$
$$+ \frac{1}{8}\sqrt{17 + 3\sqrt{17} - \sqrt{34 - 2\sqrt{17}} - 2\sqrt{34 + 2\sqrt{17}}}$$

and since this is an expression containing only square roots and numbers, it is constructible with a straightedge and compass.

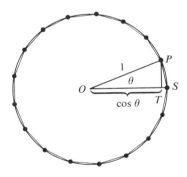

Figure 6.5

The actual construction of the 17-sided polygon is, understandably, somewhat more difficult than that of the triangle or pentagon. In fact, the method of construction was only discovered after a thorough algebraic analysis of the problem. Consider the unit circle to be drawn in the complex plane. Then the vertices of the $n$-sided polygon are $n$th roots of unity. For example, consider the 4-sided polygon, the square (fig. 6.6). The four vertices are 1, $\sqrt{-1}$, $-1$, and $-\sqrt{-1}$. All four of

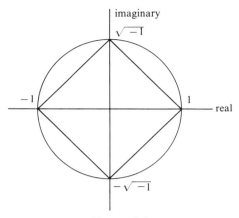

Figure 6.6

these are fourth roots of 1; they all satisfy the equation $z^4 = 1$, or $z^4 - 1 = 0$. Similarly, the vertices of the 17-sided polygon all satisfy $z^{17} - 1 = 0$ which factors into

$$(z - 1)(z^{16} + z^{15} + z^{14} + \cdots + z^3 + z^2 + z + 1) = 0$$

Gauss showed that the solution of this equation can be expressed by square roots and is, thus, constructible. He did the same for the equation

$$z^n - 1 = 0 = (z - 1)(z^{n-1} + z^{n-2} + \cdots + z + 1)$$

for the $n$-sided regular polygon, where $n = 2^{2^k} + 1$ is prime.

The converse, that no other polygons are constructible, was proved by a French bridge and highway engineer, **Pierre Louis Wantzel**, in 1837. He showed that a constructible number is the root of an irreducible algebraic equation of degree $2^m$ for $m$ equal to any nonnegative integer, and that any number which is the root of an irreducible equation not of degree $2^m$ is not constructible. Wantzel's result can be applied to prove that the "three famous problems" are impossible:

1. The side of a doubled cube satisfies $x^3 = 2$ if the volume of the given cube is one. This equation is of degree three and cannot be factored over the rational numbers, thus the side of a doubled cube cannot be constructed with ruler and compass.

2. An angle of 60° cannot be trisected with a straightedge and compass. If it could be trisected, that would mean that a 20° angle could be constructed which, in turn, would require that cos 20°

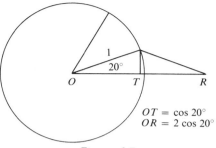

**Figure 6.7**

and 2 cos 20° both be constructible (see fig. 6.7). However, $x = 2\cos 20°$ cannot be constructed. To see this, note that

$$\frac{1}{2} = \cos 60°$$
$$= \cos 40° \cos 20° - \sin 40° \sin 20°$$
$$= (2\cos^2 20° - 1)\cos 20° - (2\sin 20° \cos 20°)\sin 20°$$
$$= 2\cos^3 20° - \cos 20° - 2(1 - \cos^2 20°)\cos 20°$$
$$= 4\cos^3 20° - 3\cos 20°$$

so that

$$1 = 8\cos^3 20° - 3\cdot 2\cos 20°$$

or

$$1 = x^3 - 3x \quad \text{where } x = 2\cos 20°$$

The equation $x^3 - 3x - 1 = 0$ is irreducible over the rationals and is of degree three. Therefore, $2\cos 20°$ is not constructible with a straightedge and compass, and, consequently, the 60° angle cannot be so trisected.

3. It was later shown by Lindemann that $\pi$ satisfies no algebraic equation and is, therefore, not constructible. Thus, the squaring of the circle is impossible.

The same day that Gauss discovered his result about polygons, he began writing his famous journal. This diary was circulated only in 1898, 43 years after Gauss died. It contained 19 pages with 146 brief statements of discoveries, the last entry being made on July 9, 1814.

Many ideas were noted in his journal which were only discovered by others many years later, such as non-Euclidean geometry and noncommutative algebra. Since Gauss did not publish often, other mathematicians had no way of knowing that he had originated ideas which they believed they were the first to discover. It is no wonder that Gauss received their ideas coolly. Gauss, a perfectionist, did not publish his ideas because he wanted to polish and develop them first. His seal was a tree with but few fruits and bore the motto, "Few, but ripe."

At the age of 21, Gauss completed a great work on the theory of numbers, *Arithmetical Research*, which he had been working on for several years. In this book he introduced the very useful congruence algebra (residue classes, or modular arithmetic). He wrote $b \equiv c$ (mod $m$) if $m$ divides $b - c$. The statement $b \equiv c$ is read, "$b$ is congruent to $c$." For example, $16 \equiv 2$ (mod 7), and $39 \equiv 7$ (4).

There are several uses of this congruence concept. We may not care about the quotient in a division, for example, but only about the remainder. Normally, when we do a division we find both the quotient and the remainder, but the algebra of congruences provides a nice method for finding the remainder only. It is much easier than doing the complete division. The method is based on the fact that if $m$ divides $b$ giving quotient $q$ and remainder $r$, then $b \equiv r$ (mod $m$), and conversely. For example, we know that if $2^{29} - 1$ is prime, then $2^{28}(2^{29} - 1)$ is a perfect number. By Fermat's theorem, to determine if $2^{29} - 1$ is prime, we need only test prime divisors of the form $2k(29) + 1 = 58k + 1$. For $k = 1$, this gives 59. It is easy to determine if 59 divides $2^{29} - 1$ by using Gauss' congruence algebra. We need only find the nonnegative number, $r$, less than 59 which satisfies the congruence $2^{29} - 1 \equiv r$ (mod 59). If $r = 0$, then 59 divides $2^{29} - 1$, while if $r \neq 0$, then 59 does not divide $2^{29} - 1$.

The steps in the solution are chosen to simplify the calculations. First choose the power of 2 nearest the modulus 59, giving $2^6 \equiv 5$ (mod 59). Square both sides to get $2^{12} \equiv 25$(mod 59). Multiply by 2, obtaining $2^{13} \equiv 50$(mod 59). Since $50 \equiv -9$(mod 59), then $2^{13} \equiv -9$(mod 59). (The only reason for this step is that $-9$ is an easier number to compute with than 50.) Square both sides again, giving $2^{26} \equiv 81 \equiv 22$ (mod 59). Multiplying by 4, $2^{28} \equiv 88 \equiv 29$(mod 59). Multiply by 2 to show that $2^{29} \equiv 58$(mod 59). Subtracting 1, $2^{29} - 1 \equiv 57$(mod 59), so that $2^{29} - 1$ is not divisible by 59. This result was found without the troublesome calculation of $2^{29}$. Other possible divisors should be tested to determine whether or not $2^{29} - 1$ is prime.

Gauss may have been led to the idea of congruences by his study of the periods of decimals.[1] He computed these periods for fractions $1/m$

---

[1]Shanks, *Solved and Unsolved Problems in Number Theory*, pp. 203–4. This book contains an interesting treatment of number theory from a historical point of view.

for $m$ up to 1000. A few fractions and their periods are listed in the following table.

| $1/m$ | $m-1$ | period length $p(m)$ |
|---|---|---|
| $1/7 = .\overline{142857}...$ | 6 | 6 |
| $1/3 = .\overline{3}...$ | 2 | 1 |
| $1/13 = .\overline{076923}$ | 12 | 6 |
| $1/37 = .\overline{027}$ | 36 | 3 |
| $1/9 = .\overline{1}$ | 8 | 1 |

The bar indicates that the sequence is repeated over and over again. Thus, $1/7 = .142857142857\ldots$. The fractions with denominators with a 2 or 5 in them were left out, because they do not produce fractions whose repeating parts begin with the first digit ($1/6 = .1\overline{6}$, $1/5 = .2\overline{0}$). Notice in the table that the period $p(m)$ is always less than or equal to $m - 1$. If $m$ is prime, it can be proved that $p(m)$ divides $m - 1$. It is not known for which primes $m$ the period $p(m)$ equals $m - 1$.

The period of a pure repeater (a fraction whose repeating part begins with the first digit) is the smallest integer $p(m)$ such that

$$10^{p(m)} \equiv 1 \pmod{m}$$

We can see this from the example of $1/37$,

```
         .027
    37)1.00000
        0
        1 00
          74
          260
          259
            1
```

The number 10 does not have remainder 1 when divided by 37 nor does 100 when divided by 37, but 1000 divided by 37 does have remainder 1. Once a remainder of 1 is obtained, the division is exactly as it was at the start of the problem, 37 into a 1 followed by a string of zeros. Thus, the first three digits of the quotient will repeat again and again. $10^3 \equiv 1 \pmod{37}$, while $10^2 \not\equiv 1 \pmod{37}$ and $10 \not\equiv 1 \pmod{37}$.

Gauss' *Arithmetical Research* was a difficult book to read, yet **Sophie Germain (1776–1831)**, a Frenchwoman, found it fascinating and sent

Gauss some of her own results on the subject, writing under the pseudonym of M. le Blanc, because she felt that a woman mathematician would not be taken seriously.[2] Her true identity was disclosed when she tried to help Gauss. The French were fighting near Gauss' home, and Sophie, knowing the general in command of those troops, wrote the general asking him to insure Gauss' safety. An emissary from the general conveyed to Gauss Sophie Germain's regard for his safety. Of course, Gauss disclaimed any knowledge of such a person and was quite surprised to learn, as he soon did, that M. le Blanc was, in fact, Sophie Germain. He admired her for having the perseverance, in spite of the prejudices and customs against a woman mathematician, to pursue her mathematical studies and to achieve good results in a very difficult field, the theory of numbers. One of her results was a partial solution to Fermat's famous last theorem. She proved that for any odd prime $p$ less than one hundred, there are no solutions to $x^p + y^p = z^p$ in integers $x$, $y$, and $z$ not divisible by $p$.

Sophie Germain became interested in mathematics after reading the story of Archimedes' death, feeling that a subject of such captivating interest to Archimedes must be worth pursuing. However, in order to study mathematics, she had to overcome strong objections from her family. Since women were not accepted as students at the École Polytechnique, Sophie instead managed to obtain the lecture notes of some professors, including Lagrange. Fortunately, Lagrange, and later Gauss, encouraged her.

Gauss was very interested in mathematical physics and made many contributions to the field. The unit of magnetism is called the gauss in his honor. In 1800 a minor planet (asteroid), Ceres, was discovered, but it was so small that it was soon lost. Gauss computed its orbit from the few observations that had been made, and he showed the observers where to look in the sky for the lost planet. Ceres was found to be where Gauss had predicted, and he became famous. Gauss was named director of the Göttingen Observatory, a position he held until the end of his life.

During the period 1821–1848, Gauss was the scientific advisor to the Hanoverian and Danish governments in a geodetic survey. He was faced, therefore, with the problems of making measurements on a curved surface, the earth. In his study he developed the idea of coordinates intrinsic to the curved surface. For example, on a sphere one can use latitude and longitude, rather than considering the sphere as part of a three-dimensional space and using the awkward $(x,y,z)$ coordinates. Intrinsic coordinates are also essential for the study of more complicated surfaces which may not be embeddable in ordinary three-dimensional

---

[2]See Lynn M. Osen, *Women in Mathematics*, pp. 83–93 for biographies of Sophie Germain and other women mathematicians.

space. Gauss showed how to express important quantities such as the length of a curve and the curvature at any point on the curve in terms of these intrinsic coordinates. The theory of surfaces is the subject of differential geometry, a study which was later applied to the general theory of relativity by Einstein.

One of the men most responsible for the development of projective geometry is **Jean-Victor Poncelet (1788–1867)**, a Frenchman. Poncelet was with the French army in Moscow and was taken prisoner. He created projective geometry while in prison in 1813–14, and his book *Treatise on the Projective Properties of Figures* was published in 1822.

Poncelet discovered the important *principle of duality*. The principle of duality states that every proposition of projective geometry remains true when the words *point* and *line* are interchanged. Thus, we get two theorems, rather than only one. For example, it is true that *two lines determine a point*, if we accept the convention that parallel lines determine a point at infinity. Interchanging point and line, we obtain the dual statement, *two points determine a line*. An interesting statement results from forming the dual of Pascal's theorem. The statement was first found by **Brianchon**. You will be asked to investigate the dual of Pascal's theorem in problem 19 of this chapter.

Another interesting concept in geometry, the one-sided surface, was discovered by **A. F. Möbius (1790–1860)**. The simple one-sided surface pictured in figure 6.8 is now called the Möbius strip. You can make a Möbius strip by taking a strip of paper, twisting it once and joining the ends together. If you place a pencil on the "top" and move it along the surface of the strip, you reach the "bottom." Continuing, you reach the "top" again without ever lifting the pencil.

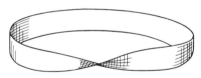

Figure 6.8

## 2 Advanced Calculus

In the mid 1700s more complicated problems of applied mathematics began to be studied. However, Bernoulli's and Euler's notions of function were not adequate for an understanding of these new problems. In the early 1800s Fourier developed new mathematical methods based on a new concept of function. The joint desires to solve new problems and to clarify the concepts of calculus led mathematicians such as Cauchy, Bolzano, Abel, and Dirichlet to develop and extend the notions of function, integral, continuity, limit, and convergence.

## JOSEPH FOURIER (1768-1830)

Fourier, a Frenchman, wrote *The Mathematical Theory of Heat* which was published in 1822. His work was not immediately accepted because it was so different from the established ideas. Fourier solved problems in heat conduction, as illustrated by the following example.

> Suppose a rectangular plate $BAC$ of infinite length (fig. 6.9) is to be heated at its base $A$, while preserving at all points of the base a constant temperature 1, and each of the two infinite sides $B$ and $C$, perpendicular to the base $A$, is kept at a constant temperature 0 at every point. What is the temperature at any point of the plate? (The solution will not be given in this text.)

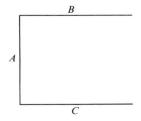

Figure 6.9

To solve problems such as this Fourier used infinite series of sines and cosines, such as

$$a_0 + a_1 \cos x + a_2 \cos 2x + a_3 \cos 3x + \cdots$$

where the coefficients $a_0, a_1, \ldots,$ are any given numbers. Fourier's claims about such series, now called Fourier series, were controversial. To understand the controversy, it is necessary to know the discoveries prior to Fourier's.

An important problem in mathematical physics, the vibrating string problem, was studied extensively about 1750.

Suppose a string is fastened at $A$ and $B$ (fig. 6.10). At time zero the string is plucked, and it vibrates. What will be the height $y(x,t)$ of the string at point $x$ and at time $t$?

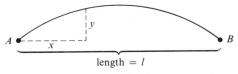

Figure 6.10

The equation formulating the law of such motion is a partial differential equation, as is the equation for the heat conduction problem.

**Jean Le Rond D'Alembert (1717–1783)** and Euler gave a general form of the solution to the vibrating string problem while **Daniel Bernoulli (1700–1782)**, the son of John Bernoulli, gave the solution as a series of sines and cosines. In Bernoulli's solution the initial position of the string at time $t = 0$ is represented as

$$y(x,0) = b_1 \sin \frac{\pi x}{l} + b_2 \sin \frac{2\pi x}{l} + b_3 \sin \frac{3\pi x}{l} + \cdots$$

where $l$ is the length of the string, $x$ is any point between $A$ and $B$, and the $b$s are constants. Bernoulli was primarily a physicist, and his arguments for such a representation were based on physical grounds. When a guitar string is plucked, for example, you first hear the main note, but as the tone fades you hear the overtones, such as the sound an octave higher. Bernoulli's sine functions are ideally suited to represent such a situation.

The function $\sin \pi x/l$ is zero at $x = 0$ and $x = l$. The function $\sin 2\pi x/l$ is zero at $x = 0$, $x = l/2$ and $x = l$. Their graphs are compared in figure 6.11. The function $\sin \pi x/l$ represents the main tone, while the

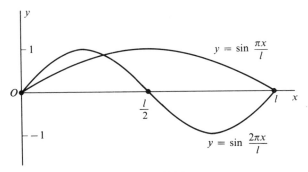

Figure 6.11

function sin $2\pi x/l$, being fixed in the middle of the string, represents a tone an octave higher (as we know from the Pythagorean studies). Similarly, sin $3\pi x/l$ is fixed at two points in between and represents an even higher tone. The tone that we hear is analyzed into its pure tone components by the ear.

Bernoulli claimed that any solution to the vibrating string problem could be represented by his trigonometric series, but Euler disagreed with him. To understand Euler's thought we must keep in mind some properties of familiar functions as Euler knew them. Functions such as $y = x^5 + 3x^2 + 1$ or $y = 1 + x + x^2/2 + x^3/6 + \cdots$ have two important properties, among others.

1. They are very smooth (fig. 6.12). No graph of such a function has corners. To represent a graph with a corner using these functions, two of them must be pieced together, such as

$$f(x) = -x \quad \text{if } x < 0$$

and

$$f(x) = x \quad \text{if } x \geq 0$$

which represents the graph with corner in figure 6.13.

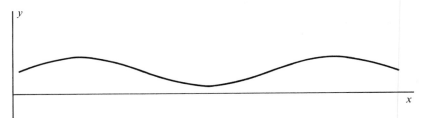

Figure 6.12

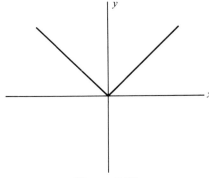

Figure 6.13

2. If two such functions agree on an interval, they agree everywhere. Thus, it is impossible to have two such functions with the following graphs (fig. 6.14). If we are more specific, a stronger statement can be made. For example, if we know only three points, we can say that there is only one circle going through those three points.

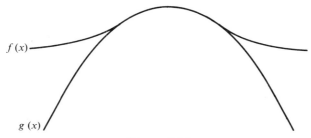

Figure 6.14

Returning to Euler's argument, note that the sine function is periodic. The function $\sin \pi x/l$ repeats itself in every interval of length $l$, just as $\sin x$ repeats itself in any interval of length $2\pi$ (fig. 6.15). Euler argued that since

$$b_1 \sin \frac{\pi x}{l} + b_2 \sin \frac{2\pi x}{l} + b_3 \sin \frac{3\pi x}{l} + \cdots$$

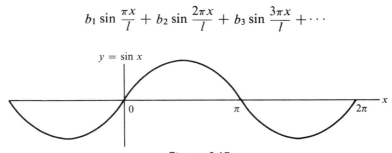

Figure 6.15

is periodic, it could not possibly represent the shape of a string which is given by a nonperiodic function. For example, consider a parabola and let the shape of the string be given by the top part of the parabola (fig. 6.16). A periodic function could not possibly have the same graph as such a string, because by property 2 above, if it agreed with the parabola on the interval $A$ to $B$, it would have to agree with it everywhere. But the parabola is clearly not periodic, thus such agreement is impossible.

Fourier in his solution of heat conduction problems used trigonometric series. He also claimed that all solutions could be represented by such series. His point, not understood by the more traditional mathema-

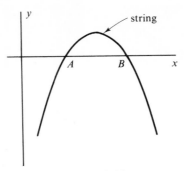

**Figure 6.16**

ticians, was that these series *are not* familiar functions. They are much more complicated than polynomials or power series like $y = 1 + x + x^2/2 + x^3/6 + \cdots$. Trigonometric series have corners; they may even have jumps, such as in fig. 6.17. A trigonometric series can represent the top of the parabola, because it has corners (see fig. 6.18).

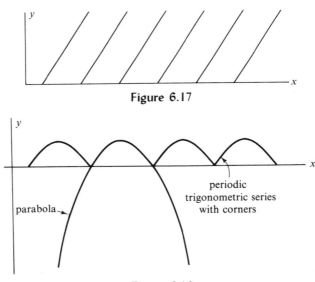

**Figure 6.17**

**Figure 6.18**

Because of Fourier's work, the entire notion of function had to be extended. A graph such as figure 6.13, which formerly was given by two functions, could now be given by one function of the new type. With the extension of the function concept, more questions arose. When is a function continuous? How can you tell from its equation whether or not it has jumps? When does a Fourier series converge? These new problems of mathematical physics required a deeper analysis of the ideas of function and limit and continuity. What we know as advanced calculus was developed in the 1800s to deal with these questions.

Problems in the theory of integration also arose from Fourier series. If a function $f(x)$ is to be represented as a series

$$f(x) = b_0 + b_1 \sin x + b_2 \sin 2x + b_3 \sin 3x + \cdots$$

how can the $b$s be found? To find $b_2$, for example, multiply the equation by $\sin 2x$ and integrate.

$$\int_0^{2\pi} f(x) \sin 2x$$
$$= \int_0^{2\pi} (b_0 \sin 2x + b_1 \sin x \sin 2x + b_2 \sin 2x \sin 2x + \cdots)$$

If term-by-term integration is valid, we get

$$\int_0^{2\pi} f(x) \sin 2x$$
$$= b_0 \int_0^{2\pi} \sin 2x + b_1 \int_0^{2\pi} \sin x \sin 2x + b_2 \int_0^{2\pi} \sin 2x \sin 2x + \cdots$$

and you can check that the entire right-hand side of the equation reduces to $2\pi b_2$, so that

$$b_2 = \frac{1}{2\pi} \int_0^{2\pi} f(x) \sin 2x$$

Thus, to find the coefficients by this method, one must evaluate an integral.

Newton made the finding of areas much easier by using antidifferentiation. Antidifferentiation works very well for functions such as $y = 1 + x^2 + x^5$, but Fourier was studying functions with corners and jumps. How can the integral be defined for such functions? Cauchy, and later Riemann, defined the integral in terms of rectangle approximations. It was then asked, which functions are integrable? Thus, Fourier series influenced the development of the theory of integration also.

## CAUCHY, BOLZANO, ABEL, AND DIRICHLET

**Augustin-Louis Cauchy (1789–1857)**, a Frenchman, published very extensively. He developed concepts in calculus and the theory of functions such as limit, continuity, and integral. He also proved many results about groups of substitutions, groups being a concept first created by Lagrange.

**Bernard Bolzano (1781–1848)**, a Czech, gave arithmetical definitions of limit, continuity, and magnitude to avoid the dependence on geometric intuition which can be misleading. This approach was not really appreciated until Weierstrass made it known some 30 to 40 years later.

**Niels Henrik Abel (1802–1829)**, a Norwegian, was eighteen when his father died, and Abel had to support his mother and six brothers and sisters. He was never in very good health, and at the age of only 26 he died from tuberculosis. Even so, he was able to make valuable contributions to mathematics.

Abel studied the developments of Newton, Euler, and Lagrange, especially the work of Newton and Euler on infinite series. Generally, in the 1600s and 1700s series were manipulated formally, just as polynomials were, and, occasionally, some strange results were obtained. For example, substituting $x = 2$ into

$$1 + x + x^2 + x^3 + \cdots = \frac{1}{1-x}$$

gives

$$1 + 2 + 4 + 8 + \cdots = -1$$

Abel, in the early 1800s, used and developed convergence criteria for series. Mathematicians were learning that one had to be more careful with infinite series than with polynomials.

Since the time of Cardano and Ferrari in 1545, general formulas for the solution of cubic and quartic equations had been known, but no one had been able to find a general solution by roots to the quintic equation. We have seen how Lagrange tried by developing the method of using groups of permutations. Abel, in 1823, used these ideas to prove that a solution by roots to the general fifth degree equation is impossible. (Paolo Ruffini, an Italian, had given a partial solution in about 1800.) Abel wrote to Gauss about his proof, but Gauss, probably thinking it was another worthless paper, did not read it. This is understandable, because mathematicians get many letters from unknowns who claim to have trisected any angle, or who claim to have proved Fermat's last "theorem." One does not expect these claims to be valid.

Abel was sent to Berlin and Paris by his government to meet the top mathematicians. Some of the established men received him coolly, but in Berlin he met August Crelle (1780–1856) an amateur mathematician who in 1826 started one of the first journals devoted exclusively to mathematical research. It is written primarily in German and continues to this day, known informally as *Crelle's Journal*. Crelle published much of Abel's research, making his work known. While in Paris, Abel gave an excellent paper on functions to Cauchy to present to the Paris academy, but Cauchy mislaid it and ultimately forgot about it.

Connected with the development of the concept of function arising from studies of Fourier series during the early 1800s is another important mathematician, **Peter Gustav Lejeune Dirichlet (1805–1859)**. Dirichlet said that $y$ is a function of $x$ if for any value of $x$ there is a rule which gives a value of $y$ corresponding to it. This idea is nearly the same as that underlying current definitions of function. Dirichlet's definition got away from the idea that a function must be given by one formula. He gave as an example the function which is zero at every rational number and one at every irrational number (fig. 6.19).

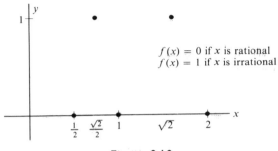

Figure 6.19

## 3 Variety in Geometry

New types of geometry were created in the early nineteenth century. Even though Euclid's geometry seemed to give a correct representation of geometry in the "real" world, mathematicians showed that it was possible to create other geometries which did not fit the pattern of reality. Mathematics was becoming free to define its own type of truth.

### NIKOLAI IVANOVITCH LOBACHEVSKY (1792-1856)

Lobachevsky, a Russian, was a bright student who became a full professor at the University of Kazan at age 23. He made his first announcement of non-Euclidean geometry there in 1826 to a mathematics society. Recall that Euclid's fifth postulate states that if two lines are cut by a transversal, and if the interior angles on the same side are less than two right angles, then the two lines intersect on that side (fig. 6.20). An equivalent form of the parallel postulate is that *given a line and a point not on the line, there is one and only one line through the point and parallel to the given line* (fig. 6.21).

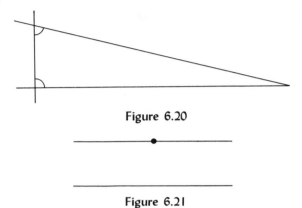

Figure 6.20

Figure 6.21

It was often argued that, even though the fifth postulate seems to be true, such a complicated statement ought to be proved and not just assumed as a postulate. Many unsuccessful attempts were made to prove the fifth postulate using the other postulates. A notable attempt was by Saccheri, who assumed the opposite of the fifth postulate and tried to deduce from it a contradiction. Lobachevsky made a breakthrough when he decided that it was impossible to prove Euclid's fifth postulate. He saw that other forms of the parallel postulate were possible and created a geometry based on the following postulate.

**Postulate**  Given any line and a point not on the line, there is at least one line through the point parallel to the given line (fig. 6.22).

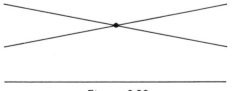

Figure 6.22

Lobachevsky's postulate appears to be false. I really do not believe that there are two parallels to a given line, through the same point not on the line. But my belief is based on a picture that I draw. This picture is a "model" for geometry. In this model I would have to say that Euclid's postulate seems true and Lobachevsky's false. However, there may be other models based on different pictures of lines in which Lobachevsky's postulate seems true and Euclid's false, while Euclid's other four postulates still seem true. To illustrate this, it is easier to consider still another geometry.

Riemann (he will be discussed later) developed a non-Euclidean geometry based on the postulate that *given any line, and a point not on it, there is no line through the point and parallel to the given line.* To develop this geometry, Riemann had to give a strict interpretation to Euclid's second postulate which states that any line can be extended indefinitely. It was usually assumed that every line is infinite. Even a closed line, such as a circle, can be extended indefinitely. In Riemann's geometry every line is finite.

A sphere can be considered as an adequate model of Riemann's geometry for our purposes. In the sphere model the word *line* is interpreted as a great circle on the sphere. A great circle is a circle whose diameter is a diameter of the sphere. In this model every two *lines* intersect (fig. 6.23); thus, there are no parallel lines. (The reason that the sphere is not a perfect model is that every two *lines* intersect in two points. This problem can be corrected, but we will ignore it.) Even though the lines on a sphere are different than on a plane, it is still true that the shortest distance between two points is along one of the lines.

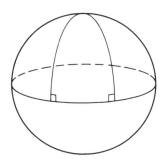

Figure 6.23

Some theorems of Riemann's geometry which are different from Euclid's are:

**Theorem 1** All perpendiculars to a line meet in one point.

**Theorem 2** The sum of the angles of a triangle is always greater than 180°, and the sum decreases to 180° as the triangle gets smaller in area.

Variety in Geometry 225

**Theorem 3**  Two similar triangles are congruent.

These theorems can be illustrated in the sphere model (fig. 6.23).

Consider some theorems of Lobachevsky's geometry. Any theorem of Euclidean geometry proved without using the idea of parallels in any way will still be true. Thus, the base angles of an isosceles triangle are still equal. A theorem different in Lobachevsky's geometry is that the sum of the angles of a triangle is always less than 180°. A model of Lobachevsky's geometry, adequate for our purposes, is pictured in figure 6.24. In this model the lines are not familiar curves such as the straight lines of the plane or the great circles of the sphere, but it is still true that the shortest distance between two points is along one of these lines, and this property can be used to sketch the lines. In figure 6.24 the lines are indicated between the three points A, B, and C. Note that the sum of the angles of the triangle ABC appears to be less than 180°.

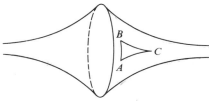

Figure 6.24

We have seen models of Riemann's geometry and of Lobachevsky's geometry in figures 6.23 and 6.24, respectively. We may have intended to study only geometry of the blackboard, in which Euclid's postulate is true. Even though Euclid's fifth postulate would then always seem true, it could not be proved from his first four postulates, because there are other models in which the first four postulates are true but the fifth is false. Thus, the truth of the first four postulates does not imply the truth of the fifth; Euclid's fifth postulate is independent of his first four. Either Euclid's, Lobachevsky's, or Riemann's postulates can be used to obtain a consistent geometry.

The discovery of non-Euclidean geometry helped to give mathematics great freedom. Many different mathematical systems may be consistent, and if so, their theory can be developed. The astounding fact is that when a new system is developed, someone will find a useful interpretation of the system. Ironically, the seemingly useless non-Euclidean geometry was used by Einstein as the basis for his general theory of relativity. "Real" space may be curved in the new systems and not flat after all. In this respect, appearances can be deceiving, because we cannot limit the real to what is easily observable from our vantage point on earth. A person who has never traveled more than 10 miles from his home may have no reason not to believe that the earth is flat, yet a longer trip would convince him of its roundness. Similarly, man may have believed the universe to be flat, only because he had experienced only a part of it.

## 4  Variety in Algebra

The creation of non-Euclidean geometry was soon followed by similar events in algebra. In the early and mid 1800s several systems of algebra were created which, like non-Euclidean geometry, used assumptions which seemed to contradict obvious truths. The first to create such a system in algebra was an Irishman, William Hamilton.

### WILLIAM ROWAN HAMILTON (1805-1865)

Hamilton was raised by his uncle who was an amateur linguist and who taught him many languages. Over a span of 10 years, beginning at age 3, Hamilton learned 13 languages, including Persian, Malay, and Bengali. In college Hamilton wrote a paper on optics and earned a high reputation. In his work on optics he formulated convenient forms of the fundamental equations of mechanics, called Hamilton's equations. At 22, while still an undergraduate, he was chosen to be a professor of astronomy, though he had not even applied for the position.

Hamilton made a significant contribution to algebra, but to discuss this we should be familiar with the development of algebra in the 1800s in England. England in the early 1800s had taken the lead in abstract algebra. Recall that English mathematics had lagged behind in the 1700s due to the hostility between English and Continental mathematicians over the Newton-Leibniz dispute. Later in the century the younger English mathematicians decided to forget the dispute and to strive for superior mathematics. Among their main interests was the abstraction of the "laws" of arithmetic and algebra. Emulating the axiomatic approach to geometry, they listed postulates, or laws, of arithmetic, such as the commutative laws of addition and multiplication. The following examples are laws of arithmetic.

$$(a + b) = (b + a) \quad \text{commutative law of addition}$$
$$ab = ba \quad \text{commutative law of multiplication}$$
$$(a + b) + c = a + (b + c) \quad \text{associative law of addition}$$
$$(ab)c = a(bc) \quad \text{associative law of multiplication}$$
$$a(b + c) = ab + ac \quad \text{distributive law}$$

**George Peacock (1791–1858)** in 1830 published *Treatise on Algebra* in which he applied the commutative and associative properties of numbers to symbolic algebra. He and **Augustus De Morgan (1806–1871)** noted that the above laws apply to numbers, to algebra, and to geometrical magnitudes.

In 1833 Hamilton showed how to consider a complex number $a + b\sqrt{-1}$ as a pair of real magnitudes, $(a,b)$. He showed that one could work with the pair $(a,b)$, never mentioning $\sqrt{-1}$, and still obtain the same results as if $\sqrt{-1}$ had been used. Hamilton performed addition and multiplication of two complex numbers $a + b\sqrt{-1}$ and $c + d\sqrt{-1}$. For example,

$$(a + b\sqrt{-1}) + (c + d\sqrt{-1}) = (a + c) + (b + d)\sqrt{-1}$$

and

$$(a + b\sqrt{-1})(c + d\sqrt{-1}) = ac + ad\sqrt{-1} + bc\sqrt{-1} - bd$$
$$= (ac - bd) + (ad + bc)\sqrt{-1}$$

He noted that he did not really need $\sqrt{-1}$ to express these calculations. They could be written

$$(a,b) + (c,d) = (a + c, b + d) \tag{1}$$
$$(a,b)(c,d) = (ac - bd, ad + bc) \tag{2}$$

with the understanding that the second component always is multiplied by $\sqrt{-1}$ and added to the first.

Hamilton also had the insight to reverse the process. He defined a complex number as a pair of real numbers $(a,b)$ where addition and multiplication was performed according to the definitions in equations (1) and (2). Based on these definitions, he was able to prove that the familiar laws such as the associative, commutative, and distributive laws were true for complex numbers. Thus, Hamilton reduced the mysterious complex numbers to pairs of real magnitudes for which addition and multiplication satisfied the associative, commutative, and distributive laws.

Hamilton was quite interested in physics, and because of his interest he found Wessel's representation of complex numbers as rotations useful. He wanted to develop a similar algebra of triples $(a,b,c)$ to obtain an algebra of rotations in three dimensions, but he never could find a way to define the multiplication $(a,b,c) \cdot (d,e,f)$, though he spent many years trying. Finally, in 1843 he developed an algebra of four quantities $(a,b,c,d)$ that he called *quaternion algebra*. However, this algebra did not satisfy all the laws. Hamilton had to break the commutative law of multiplication which was quite a radical step as all the previous studies of the laws of integers, geometric magnitudes, and algebra led algebraists to the conclusion that the laws always held.

In creating non-Euclidean geometry, Lobachevsky denied Euclid's parallel postulate which seemed true; in creating quaternion algebra, Hamilton denied the commutative law which seemed true. The point is that Euclid's postulate may very well be true in the usual model of geometry, but there are other models of geometry for which it is false. Even though all the systems of arithmetic and algebra studied up until 1843 satisfied the commutative law, Hamilton found a system which did not. Hamilton showed that in algebra, too, one was free to develop any logically sound system.

Hamilton was quite impressed with his quaternions and spent the rest of his life developing their properties. He tried to relate three components of $(a,b,c,d)$ to rotations in three dimensions. Later, in 1881, an American, **J. W. Gibbs (1839–1903),** developed vector algebra based on Hamilton's work.

## ARTHUR CAYLEY (1821-1895)

Cayley was a top English mathematician who wrote extensively in algebra and other fields of mathematics, including transformations of variables. Cayley developed another type of algebra, the algebra of matrices. You may recall that it is useful sometimes to change coordinates, for example to simplify the equation of an ellipse. We are given $x$ and $y$ axes and wish to change these to $x'$ and $y'$ axes (fig. 6.25). If the transformation is linear, we may express it as

$$x' = ax + by$$
$$y' = cx + dy$$

Suppose we make another change from $x'$ and $y'$ axes to $x''$ and $y''$ axes with equations

$$x'' = ex' + fy'$$
$$y'' = gx' + hy'$$

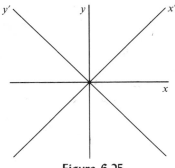

Figure 6.25

To find the equations of the change from $x$ and $y$ axes to $x''$ and $y''$ axes we substitute, giving

$$x'' = e(ax + by) + f(cx + dy)$$
$$y'' = g(ax + by) + h(cx + dy)$$

or

$$x'' = (ea + fc)x + (eb + fd)y$$
$$y'' = (ga + hc)x + (gb + hd)y$$

Cayley realized that the variables themselves were not important to the calculation, only the coefficients, so he wrote the coefficients as a matrix. The matrix of the first change is $\begin{bmatrix} a & b \\ c & d \end{bmatrix}$, and that of the second is $\begin{bmatrix} e & f \\ g & h \end{bmatrix}$. Cayley defined the product of these two matrices to be the result of successive transformations. Thus, the product matrix would give the transformation from the $x$ and $y$ axes to the $x''$ and $y''$ axes, and

**Definition**

$$\begin{bmatrix} e & f \\ g & h \end{bmatrix} \begin{bmatrix} a & b \\ c & d \end{bmatrix} = \begin{bmatrix} ea + fc & eb + fd \\ ga + hc & gb + hd \end{bmatrix}$$

The complicated definition of matrix multiplication that one learns was chosen because it represents a naturally-occuring and useful situation, that of a transformation of variables.

The commutative law does not hold in matrix multiplication. Thus, here again, an algebra was defined which does not obey all the laws of the arithmetic of fractions. As an example,

$$\begin{bmatrix} 2 & 3 \\ -1 & 0 \end{bmatrix} \begin{bmatrix} 1 & 0 \\ -1 & 2 \end{bmatrix} = \begin{bmatrix} -1 & 6 \\ -1 & 0 \end{bmatrix}$$

while

$$\begin{bmatrix} 1 & 0 \\ -1 & 2 \end{bmatrix} \begin{bmatrix} 2 & 3 \\ -1 & 0 \end{bmatrix} = \begin{bmatrix} 2 & 3 \\ -4 & -3 \end{bmatrix}$$

### EVARISTE GALOIS (1811-1832)

Evariste Galois, a Frenchman, was a most unlucky man. He had problems in school, because he was so bright that his teachers could not understand him. The technical school examiners could not understand his calculations on his entrance examination, either. They failed him twice, because he used his own method of computation, rather than following the method suggested by the examiners. Galois' bad luck continued as he grew older. He submitted a paper containing his original work in algebra to the French Academy of Sciences, but the secretary, Fourier, passed away before he could review it. Fourier's successor could not understand Galois' paper, so he returned it to the luckless mathematician. To compound his misfortunes, Galois was jailed for his political activities. Soon after his release, he became involved in a duel. After spending the previous night writing his discoveries, Galois died in the duel at the age of 21.

Galois' mathematical discoveries were brilliant. We have seen how Lagrange used groups of permutations in trying to solve equations, and that Abel proved that no solution by roots exists for the general fifth degree equation. This does not mean that all fifth degree equations cannot be solved by roots, but that there is no general formula for the solution analogous to the quadratic formula. For example, a solution of $x^5 = 2$ is $\sqrt[5]{2}$. Lagrange took the roots $x_1$, $x_2$, $x_3$, $x_4$, $x_5$ of the general quintic and looked at the group of their 120 permutations. By contrast, Galois defined a special group for each equation. He then gave properties of this group which were equivalent to the solvability of the equation by roots. Thus, an equation like $x^5 + 4x^2 + 1 = 0$ would have its own Galois group, and the properties of this group would determine whether or not this equation has a solution by roots. After Galois' death other mathematicians worked out the implications of his theory.

# GEORGE BOOLE (1815-1864)

Boole, an Englishman, spent four years teaching elementary school, then at age 20 he opened a school of his own. As part of the instruction, Boole taught his students mathematics. He became very interested in the subject as a result and studied some very advanced works. Soon Boole became an accomplished mathematician in his own right.

Recall that the English mathematicians had taken the lead in expressing the "laws" of algebra. Boole, in *Mathematical Analysis of Logic* in 1847 and *Investigation of the Laws of Thought* in 1854, clearly stated the view that mathematics involves symbols and their rules of combination subject only to inner consistency. Boole developed an algebra of classes which can be applied to derive laws of correct reasoning and is now the subject of symbolic logic. His algebra of classes, or sets, is called *Boolean algebra*, and applications of Boolean algebra recently have been made to the logical design of circuits in computers.

The operations of interest for sets are union and intersection. The union of two sets $a$ and $b$, $a \cup b$, is defined as the set containing all elements which are either in set $a$ or in set $b$ or in both. The intersection of sets $a$ and $b$, $a \cap b$, is the set containing all elements which are in both sets $a$ and $b$. Some of the rules for Boolean algebra are analogous to those for arithmetic, such as $a \cup b = b \cup a$ and $a \cap b = b \cap a$, which compare with $a + b = b + a$ and $ab = ba$, respectively, the commutative laws. A rule which is different is $a \cup (b \cap c) = (a \cup b) \cap (a \cup c)$. The arithmetic analogue would be $a + bc = (a + b)(a + c)$ which is false. Also in Boolean algebra, $a \cup a = a$ and $a \cap a = a$, while in arithmetic it is not a law that $a + a = a$ or $a \cdot a = a$. Thus, Boole created an algebra of classes which has its own system of laws.

De Morgan also wrote on symbolic algebra. A student of his posed a problem which has since become a famous one. What is the least number of colors needed in order to be able to color any map? Each region is to be colored in such a way that no two bordering regions, unless the regions meet at isolated points, have the same color. Any map can be colored with five colors. Some maps require four colors and cannot be colored with any less than four colors. No one has ever found a map that needs five, rather than four, colors, but, at the same time, no one has yet been

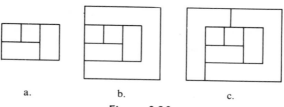

a.        b.        c.

Figure 6.26

able to prove that four colors are sufficient to color any map. (Figure 6.26 contains three examples that you might color to get a feeling for the problem.)

# 5 The Arithmetization of Analysis

In the early 1800s, based on the research of Fourier, the concepts of calculus had become more complex. Visual intuition could no longer be trusted to give correct results; mathematicians sought a more secure foundation for the calculus. In this effort to build a solid base, the modern concepts of limit and real number were created.

### KARL WEIERSTRASS (1815-1897)

Weierstrass, a German, was sent to law school by his father, but he was more interested in fencing and drinking beer than in law. He came home disgraced, without a degree. Weierstrass did pass the state examination for teachers, however, and he taught school for 15 years, doing research in mathematics at night. He was particularly interested in Abel's work. In 1854 Weierstrass had a paper printed in *Crelle's Journal*, and he became famous. Even though he was teaching in a small town of which few mathematicians in Berlin had ever heard, he was doing work comparable to that of any mathematician and was, therefore, invited to teach at Berlin. Because of all the excitement of his quick rise to fame, Weierstrass suffered a nervous breakdown. From that time on he experienced dizzy spells when standing, so he remained seated while teaching and had his students write formulas on the board.[3]

Weierstrass was a very clear thinker, and he noticed fine points in reasoning which had escaped previous mathematicians. Perhaps the most familiar contribution of Weierstrass would be the $\epsilon$-$\delta$ definition of limit encountered in calculus courses. The basic concepts of calculus were none too clear in the exposition of Newton and Leibniz. Thereafter, there was much discussion of limit concepts, and several definitions were given. The earlier writers were able, nevertheless, to apply calculus to polynomial and power series functions.

---

[3]See Bell, *Men of Mathematics*, pp. 406-32, for a biography of Weierstrass.

As was mentioned in connection with Fourier, new ideas about functions (functions with corners, discontinuous functions) were needed in order to solve partial differential equations (fig. 6.27). In the mid 1800s it was considered geometrically obvious that a continuous function could have a finite number of corners, or even infinitely many corners, but that between these corners it was smooth and had a well-defined tangent or derivative. It seemed obvious, and was so stated, that a continuous function has a derivative at almost all its points, even though it may have many corners (fig. 6.28).

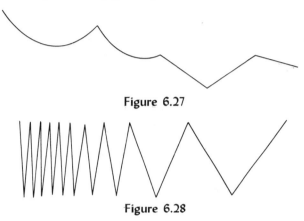

Figure 6.27

Figure 6.28

There was a certain amount of uneasiness among mathematicians about basing results on geometric intuition. Weierstrass justified this feeling by showing that the above "geometrically-obvious" statement that a continuous function has a derivative at nearly every point is wrong. He gave an example of a function which is continuous but nowhere differentiable. It has corners everywhere! Such a function is impossible to sketch, yet it can be proved to exist.

By the later nineteenth century some mathematical concepts were so complex that they could not be understood from a diagram. In fact, a picture might give a misleading impression, for example, that a continuous function seems to be smooth at most points. Mathematicians felt that they must avoid making geometrically-obvious statements, and base their reasoning on sound principles of arithmetic. This program is sometimes called the *arithmetization of analysis*.

Bolzano, in 1830, was the first to give an example of a continuous nowhere differentiable function, but the impact of this discovery was only felt in 1861 with the example given by Weierstrass. In his lectures Weierstrass presented the function

$$f(x) = \sum_{n=0}^{\infty} b^n \cos(a^n x \pi)$$

where $a$ is an odd integer greater than one, and $b$ is a positive constant less than one. He proved that this function is continuous and nowhere differentiable. We will study the arithmetization of analysis again in connection with Dedekind.

## BERNHARD RIEMANN (1826-1866)

Riemann, another German mathematician, was a man of great insight. He formulated ideas which have required generations of mathematicians to develop. While still in high school, Riemann mastered a long and advanced work in number theory in six days, an accomplishment which probably inspired his later work in prime number theory. The prime numbers — 2, 3, 5, 7, 11, 13, 17, ... — have a very irregular distribution. Many cases have been found of primes which differ by only two, such as 11 and 13, but there are also arbitrarily large intervals between primes. Although no formula has been found to give the exact number of primes less than a given number $x$, approximate results can be found. For large numbers $x$, the number of primes, $\pi(x)$, less than $x$ is approximately $x/\log x$. Riemann was unable to prove this result completely. In his study of the distribution of primes he used the *Riemann zeta function*

$$\zeta(s) = 1 + \frac{1}{2^s} + \frac{1}{3^s} + \frac{1}{4^s} + \frac{1}{5^s} + \cdots$$

where $s$ is any complex number $u + iv$.

One of the most famous unsolved problems in mathematics is concerned with the location of the zeros of the zeta function. The *Riemann hypothesis* is that all zeros of $\zeta(s)$ with $u$ between 0 and 1 are of the form $1/2 + iv$. This conjecture has neither been proved nor disproved.

The *prime number theorem*, stating that the number of primes, $\pi(x)$, less than $x$ is approximately $x/\log x$, or more precisely,

$$\lim_{x \to \infty} \frac{\pi(x)}{x/\log x} = 1$$

was finally proved in 1896 by **C. J. de la Vallée-Poussin** and **Jacques Hadamard** who arrived at their conclusions independently. When I was in college, my number theory teacher told me of a conjecture that anyone who proved the prime number theorem was immortal. At that time both de la Vallée-Poussin and Hadamard were very old. However, neither man proved to be immortal as de la Vallée-Poussin died at age 95 and Hadamard at age 97.

In his study of Fourier series, Riemann added to Cauchy's work and developed the definition of what is now called the *Riemann integral*, the type of integral studied in calculus. Riemann returned to the idea of rectangle approximations to integrate the complicated functions arising from Fourier series, because not all integrations could be reduced to antidifferentiation using rules such as $\int x^n = x^{n+1}/(n+1)$.

In his study of functions of a complex variable such as $\sqrt{z}$, Riemann gave impetus to the subject of the topology of surfaces, and he created what are now known as *Riemann surfaces*. (Much of Riemann's work in this area is so advanced that it is only studied by graduate students and professional mathematicians.) The function $\sqrt{z}$ is two-valued; every complex number has two square roots, for example, $\sqrt{4} = \pm 2$. We can also find the two square roots of $\sqrt{-1}$. Recall that if $\sqrt{-1} = (b)(b)$, then the angle of $b$ must be 45° or 225°, and its length must be 1. Thus, $b = \pm\sqrt{2}/2 + (\sqrt{2}/2)\sqrt{-1}$ (fig. 6.29). The function $\sqrt{z}$ can be thought of as a mapping from one complex plane to another, under which mapping every point in the first plane is mapped to two points in the second (fig. 6.30).

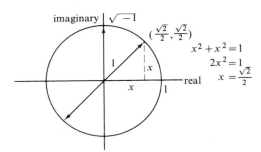

Figure 6.29

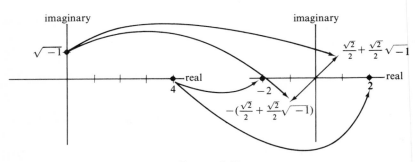

Figure 6.30

Riemann had the clever idea of avoiding this two-valuedness by doubling the domain. Instead of considering $\sqrt{z}$ as a mapping from one plane to another plane, he defined $\sqrt{z}$ as a mapping from a Riemann surface to another plane. The Riemann surface is basically two planes joined together in such a way as to form one surface (fig. 6.31). On the upper plane (shown in cross section) the square root of a number is positive; on the lower it is negative. Thus, the square-root function is single-valued if it is defined on the Riemann surface. The surface is constructed in such a way that it is impossible to make a complete loop around the origin and stay entirely on the top plane or entirely on the bottom plane. If one starts at 4 on the top plane and circles counterclockwise reaching $-4$, as he continues to circle, he will go down the "ramp" and reach the point 4 again on the lower plane. If another complete circle is made, he will reach 4 again, but this time on the upper plane. Riemann had quite an imagination — this is only the simplest Riemann surface.

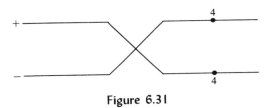

Figure 6.31

To obtain the right to lecture at the University of Göttingen without salary while waiting for a salaried position, Riemann had to present a lecture to the faculty which included Riemann's former teacher, Gauss. Riemann submitted three topics from which the faculty could choose. Gauss chose the third on which Riemann was least prepared, but which most interested Gauss. Riemann spoke "On the hypotheses which lie at the foundations of geometry." He considered a very general case where the geometry of "space" could change from point to point. The geometry might be Euclidean near one point, but Lobachevskian near another, and so on. He could also represent the geometry of a surface which was flat in parts, spherical in other parts, and so on. This is the subject of differential geometry which is, as you might guess, much more complicated than high-school geometry. Here again, mathematicians were kept busy developing the consequences of Riemann's insights. Einstein used these general geometries in his theory of relativity.

## RICHARD DEDEKIND (1831-1916)

As mentioned before, there developed among mathematicians an aversion to reliance on geometric intuition. Another example of the reliance on geometric intuition was the statement of the intermediate value theorem — a continuous function on the interval [a,b] which is positive at a and negative at b must be zero for some point x between a and b. This theorem seems obvious, but are we deceived, as in the case of a nowhere differentiable continuous function? Can the graph somehow be continuous and still jump across the x-axis (fig. 6.32)?

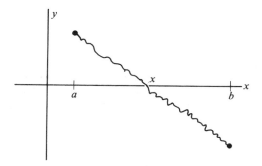

Figure 6.32

Until the mid 1800s, the x and y variables were thought of as continuous magnitudes. It was understood that the laws of arithmetic applied to magnitudes, thus, $x_1 + x_2 = x_2 + x_1$, and so on. However, the continuity property of the line of magnitudes was never analyzed clearly until Richard Dedekind, a German mathematician and the last student of Gauss, did so. Dedekind published his work *Continuity and Irrational Numbers* in 1872, though it had been written in 1858. In it he says,

> As professor in the Polytechnic School in Zurich I found myself for the first time obliged to lecture upon the elements of the differential calculus and felt more keenly than ever before the lack of a really scientific foundation for arithmetic.... For myself this feeling of dissatisfaction was so overpowering that I made the fixed resolve to keep meditating on the question till I should find a purely arithmetic and perfectly rigorous foundation for the principles of infinitesimal analysis. The statement is so frequently made that the differential calculus deals with continuous magnitude, and yet an explanation of this continuity is nowhere given ....[4]

In analyzing the continuity of the line, Dedekind compared the rational numbers with the points on the line. He gave three properties of the rationals and three analogous properties of the points on a line.

---

[4]Richard Dedekind, *Theory of Numbers*, pp. 1–2.

| Rationals (fractions) | Points |
|---|---|
| 1. Given three rationals $a$, $b$, $c$, if $a > b$, and $b > c$, then $a > c$. | 1. Given three points, if $a$ is to the right of $b$, and $b$ is to the right of $c$, then $a$ is to the right of $c$. |
| 2. Given rationals $a$ and $c$, there are an infinite number of rationals between them. | 2. Given points $a$ and $c$, there are an infinite number of points between them. |
| 3. A number $a$ divides the rationals into two classes — the class $A_1$ of rationals, such that if $a_1$ is in $A_1$, $a_1 < a$; and the class $A_2$ of rationals, such that if $a_2$ is in $A_2$, then $a_2 > a$. The number $a$ can be put in either $A_1$ or $A_2$. | 3. A point $p$ divides the points on the line into two classes — the class $P_1$ of points to the left of $p$, and the class $P_2$ of points to the right of $P$. |

Dedekind continued, "The analogy becomes a correspondence when we pick an origin and a unit length on the line.... Of greatest importance, however, is the fact that in the straight line there are infinitely many points which correspond to no rational number [fig. 6.33]."[5] There is a point on the line whose distance from the origin is just the length of the diagonal of a square of side one, $\sqrt{2}$. But we know that $\sqrt{2}$ is not a rational number. Thus, points such as $\sqrt{2}$ and $\pi$ on the line have no corresponding rational number.

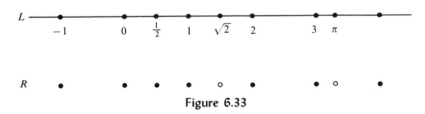

Figure 6.33

"If now, as is our desire, we try to follow up arithmetically all phenomena in the straight line, the domain of rational numbers is insufficient and it becomes absolutely necessary that the instrument $R$ constructed by the creation of the rational numbers be essentially improved by the creation of new numbers such that the domain of numbers shall gain

---

[5]Ibid., pp. 7-8.

the same completeness, or as we may say at once the same *continuity*, as the straight line. . . . In what does this continuity consist? . . . The majority may find its substance commonplace."⁶

Dedekind concluded that the essence of the continuity of the line is the converse of property 3 above concerning points. "If all points of the straight line fall into two classes such that every point of the first class lies to the left of every point of the second class, then there exists one and only one point which produces this division of all points into two classes . . . ."⁷ Thus, if a line is cut into two parts, there is exactly one point that is cut through. In contrast, it is possible to cut the rational numbers into two classes without cutting a rational number. For example, we can divide the rationals into all $a/b$ such that $(a/b)^2 < 2$, and all $a/b$ such that $(a/b)^2 > 2$. In figure 6.34a the cut goes through the point $\sqrt{2}$. In the corresponding cut in the rationals (fig. 6.34b) there is only a space where $\sqrt{2}$ is on the line, so the rationals can be divided into two classes without there being a number between the classes.

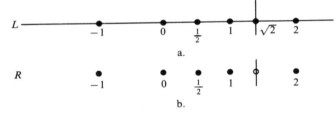

Figure 6.34

Dedekind's achievement was to explicitly state this continuity property of the line which may be taken as an axiom about the line. Then all other statements such as the intermediate value theorem may be proved using it and the other already designated properties of the line. Dedekind also saw that he could dispense with the line entirely, since the set of all cuts in the rationals has the same properties as the line, including continuity. Dedekind, therefore, achieved his goal of finding an arithmetic foundation for analysis.

Mathematics has long held the fascination of people not involved in the sciences, as well as that of professional mathematicians. President **James A. Garfield (1831–1881)** developed an original proof for the Pythagorean theorem. **C. L. Dodgson (1832–1898)**, better known as Lewis Carroll, was a mathematician by profession, although he earned his fame as the author of *Alice in Wonderland*. Among his mathematical writings is a collection of mathematical recreations.

---

⁶Ibid., pp. 9, 11.
⁷Ibid., p. 11.

# GEORG CANTOR (1845-1918)

When the concept of infinity had arisen in mathematics it had generally been rejected as paradoxical. Eudoxus and Archimedes used the method of exhaustion which avoided breaking figures into infinitely many indivisibles. The limit method, the founding concept of calculus, triumphed while methods using indivisibles were abandoned. Cantor, a German mathematician of the late nineteenth century, was well acquainted with medieval views on the infinite, and he accepted infinity as a valuable concept.

Cantor encountered infinite sets in his early studies of trigonometric series when he was seeking conditions which would insure that a function had only one trigonometric series representation. He found that the series did not have to converge to the function at every point of the interval under consideration for his theorem to be true; he could exclude any finite set and some infinite sets. Cantor found that he was studying more and more complicated types of infinite sets. Some paradoxical properties of infinite sets were expressed by Cantor, but rather than rejecting infinite sets, he pressed on with his study.

Cantor showed that a small segment of a line has just as many points, in a sense, as an infinite line. Consider a semicircle, and a straight line under it (fig. 6.35). To every point on the circle one point on the line

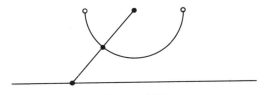

Figure 6.35

can be made to correspond, and vice versa. This one-to-one correspondence can be set up by drawing lines from the center of the semicircle through the points on the semicircle. Such lines intersect both the semicircle and the infinite line at corresponding points. The whole line is not greater than the part, at least in the number of points that each contains. This is a characteristic property of infinite sets. The whole can have a one-to-one correspondence with a proper part. A simpler example was given by Galileo, who corresponded the integers to their squares

    1    2    3    4    5    6    ...

    1    4    9   16   25   36   ...

This principle of infinite sets has some amusing applications. Suppose you have a hotel with an infinite number of rooms which are all full, and a new guest requests a room. You can easily accommodate him by moving the person in room 1 to room 2, the person in room 2 to room 3, and so on, and then giving the new guest room 1 which is now empty.

Cantor gave an example of a set, the *Cantor set*, which illustrates the difficulties of set theory. This set is difficult to visualize, and it has many properties which at first seem strange. To construct the Cantor set, start with the interval [0, 1]. Remove the middle third, namely (1/3, 2/3). Then from each of the two remaining pieces, remove the middle third (fig. 6.36). From each of the four remaining pieces remove the middle third. Continue on in this manner. The pieces remaining get continually smaller in length with each removal, but each time there are twice as many of them. After repeating this process an infinite number of times, there are some points still not removed. These remaining points form the Cantor set. The points 1/3 and 2/3 are in the Cantor set, among others. It can be shown by summing a geometric series that the intervals removed from [0, 1] have total length 1. Thus, the Cantor set has total length zero. Yet it can also be shown that there are just as many points in the Cantor set as there are in the whole interval [0, 1]!

It is no wonder that Cantor became ill and died in a mental hospital after a nervous breakdown.

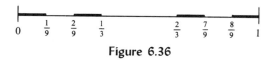

**Figure 6.36**

You might begin to believe that any one infinite set has as many points as any other. But Cantor showed this was not true by proving that points on a line cannot be in one-to-one correspondence with the integers. There are also sets with more elements than the set of points on the line.

In addition to comparing the sizes, or cardinalities, of sets, Cantor also considered the order types of sets. Consider the set $\{1/2, 2/3, 3/4, 4/5, 5/6, \ldots, 3/2, 5/3, 7/4\}$. If we were counting the elements in this set in order from smallest to largest, we would say $1, 2, 3, 4, \ldots, \omega + 1, \omega + 2, \omega + 3$. We count up to infinity and then continue further with $\omega$ representing the first infinity. The set

$$\left\{\frac{1}{2}, \frac{2}{3}, \frac{3}{4}, \ldots, \frac{3}{2}, \frac{5}{3}, \frac{7}{4}, \frac{9}{5}, \ldots\right\}$$

is of order type $\omega + \omega$. One can also define sets which have even higher order types.

It is not surprising that some mathematicians rejected Cantor's work on sets, even though many of them tried to understand his perplexing sets. His work differed so greatly from the then conventional mathematics that some mathematicians were not convinced of its validity. Most mathematicians now realize the usefulness and validity of Cantor's ideas, however, there remain some who are not convinced. The mathematicians of the time are the people who determine what is or is not valid mathematics. Many new ideas arose in the course of the history of mathematics that were not accepted immediately, but eventually proved useful. Others were accepted and then later discarded in favor of yet different ideas. Mathematics is a continually evolving science. Cantor called mathematics a *free creation*. This description expressed a view of mathematics that was coming to the forefront at the onset of the twentieth century.

# 6 The Generality of Mathematics in the Twentieth Century

A great variety of mathematical systems were created in the 1800s. By the turn of the century, this experience with specific axiomatic theories was being used in discussions of the nature of axiomatic systems. Mathematicians were beginning to understand that one axiomatic system could apply to many different models. Such more abstract axiomatic systems have been studied extensively in the twentieth century. An economy of effort is effected in this way, as one general theory may eliminate the need for many special theories which the more general one encompasses. Often a familiar subject and an unfamiliar one are seen to be special cases of the same general theory. By means of this analogy, the familiar subject can be used profitably to study the unfamiliar subject.

## DAVID HILBERT (1862-1943)

Hilbert was a German mathematician who did fundamental work in number theory, algebra, logic, calculus of variations, mathematical physics, and the foundations of geometry. He was a methodical thinker, and when he studied a subject he penetrated to its roots. In 1899 Hilbert published *Foundations of Geometry* in which he gave a clear, axiomatic system for geometry. The words line, point, and plane had no meaning except as expressed in the axioms, but they could be given any interpretation which made the axioms true. This formal axiomatic approach crystallized the developments of the nineteenth century and served as a guide to twentieth century mathematicians. Appropriately, it appeared just at the turn of the century.

In his axioms Hilbert expressed properties of concepts such as betweenness which, apparently, Euclid had taken for granted as intuitively obvious. The notion of betweenness refers to points on a line. For example, given three distinct points on a line, there is exactly one point which is between the other two. Hilbert's axioms (for geometry of three dimensions) were expressed in five groups.

I. Axioms of Connection
These seven axioms establish a connection between points, lines, and planes. For example, two distinct points $A$ and $B$ always determine a line.

II. Axioms of Order
The five axioms of order define the idea expressed by the word *between*.

III. Axiom of Parallels

IV. Six Axioms of Congruence

V. Axiom of Continuity

By giving up fixed interpretations of such concepts as point and line, mathematics may seem to be renouncing its substance. Yet this approach allows mathematics the freedom to be about anything. Thus, one theory may apply to many diverse subjects in which a similar structure is encountered, rather than apply to only one subject. We will investigate several examples of such general structures which were developed extensively in the twentieth century and which have many diverse applications. Numerous other examples also exist.

## METRIC SPACE AND THE RING STRUCTURE

Many different geometries, and other subjects, can be unified under the general concept of a *metric space*. This concept was developed in the early 1900s by **Maurice Fréchet (1878–1973)** and **Felix Hausdorff (1868–1942)**. A metric space is a set of *things* (of no specific nature), together with a distance function giving the distance between any two things $a$ and $b$. The distance function must satisfy the following properties which hold for ordinary distance between two points.

$d(a,b) \geq 0$    The distance is always greater than or equal to zero.

$d(a,a) = 0$    The distance from a thing to itself is zero.

$d(a,b) = d(b,a)$    The distance from $a$ to $b$ equals the distance from $b$ to $a$.

$d(a,b) + d(b,c) \geq d(a,c)$    The direct distance from $a$ to $c$ is less than the indirect distance via $b$.

Of course, this distance applies to ordinary plane geometry in which the distance between two points, $a = (a_1, a_2)$ and $b = (b_1, b_2)$, may be represented by the formula

$$d(a,b) = \sqrt{(a_1 - b_1)^2 + (a_2 - b_2)^2}$$

It is very useful to consider higher dimensional geometry, but it is impossible to visualize such a thing as five-dimensional space. Yet if we consider five-tuples of real numbers, $(a_1, a_2, a_3, a_4, a_5)$, as our set of things and give distance by the formula

$$d(a,b) = \sqrt{(a_1 - b_1)^2 + (a_2 - b_2)^2 + (a_3 - b_3)^2 + (a_4 - b_4)^2 + (a_5 - b_5)^2}$$

we get another example of a metric space. Many concepts from two-dimensional space carry over by analogy to five-dimensional space. Five-dimensional geometry is more useful when applied to the study of systems of linear equations, however, than as a course of study in itself.

The concept of a metric space can be applied to functions. If we let the set of things be the continuous functions on the interval [0, 1], we can define a distance function by

$$d(f,g) = \int_0^1 |f(x) - g(x)| \, dx$$

Since this distance function satisfies the same properties as ordinary distance in the plane, any theorem based on the distance axioms will have a true interpretation in both systems. We can better understand a theorem about the abstract class of continuous functions by considering its analogue in the plane of points. Finding a common metric space structure in diverse systems allows us to understand an abstract unintuitive system — functions — by means of our knowledge of an analogous, more concrete system — points.

Consider an abstract structure of algebra, the *ring*. A ring is a set with two operations, ∘ and ∗, such that

1. $(a \circ b) \circ c = a \circ (b \circ c)$
2. $a \circ b = b \circ a$
3. There is a $u$ such that $u \circ a = a \circ u = a$
4. For all $a$, there is an $a^{-1}$ such that $a \circ a^{-1} = a^{-1} \circ a = u$
5. $(a * b) * c = a * (b * c)$
6. $a * (b \circ c) = (a * b) \circ (a * c)$

Note that property 3 asserts the existence of an element $u$ which behaves like 0, when $\circ$ is ordinary addition and the elements are numbers. We know that $0 + a = a + 0 = a$.

The concept of a ring was developed by **Ernst Kummer (1810–1893)** and Dedekind. Many theorems about such structures have been proved in the twentieth century. In particular, **Emmy Noether (1882–1935)** incorporated earlier results into the abstract theory while developing it much further. The ring structure is common to many diverse systems, and all ring theorems can be applied and interpreted in each system. This common use of the theorems gives rise to a great economy, in that once a theorem is proved for the general ring structure, it need not be proved in each separate system. One of the goals of the "New Math" is to show that mathematics deals with structures such as the ring.

The following examples are a few illustrations of systems to which the ring structure applies.

**Example 1** The set of integers $0, \pm 1, \pm 2, \pm 3, \ldots$ where $\circ$ is addition and $*$ is multiplication.

**Example 2** The set 0, 1, 2, 3 where addition and multiplication are performed by reducing to remainders obtained on division by 4. Thus, $2 + 3 = 5 \equiv 1$, and $2 \cdot 3 = 6 \equiv 2$. Using the $\circ$ and $*$ notation, we would write $2 \circ 3 = 1$ and $2 * 3 = 2$.

**Example 3** Matrix algebra where the set of things is the set of $2 \times 2$ matrices such as $\begin{bmatrix} 1 & 2 \\ 0 & -1 \end{bmatrix}$ or $\begin{bmatrix} 3 & -2 \\ 4 & 1 \end{bmatrix}$. The operation $\circ$ is matrix addition, and $*$ is matrix multiplication.

**Example 4** The set of things is the set of real-valued functions $f, g, h, \ldots$ of a real number. Here $f \circ g$ is the function given by

$$(f \circ g)(x) = f(x) + g(x)$$

and $f * g$ is the function given by

$$(f * g)(x) = f(x)g(x)$$

An influential continuing publication on "structure," *Elements of Mathematics*, is a series of books which was started in 1939 by a group of young French mathematicians writing under the pseudonym of *N. Bourbaki*. (Charles Bourbaki was a French general who lived from 1816 to 1897, and there is a statue of him in the city of Nancy where many of these mathematicians lived.) The initial $N$ denotes the arbitrary number $n$ of mathematicians composing the group. The series of books is still being written, with volumes published only after extensive discussion. Members must retire from the group at age 50 and thereafter publish under their own names. The pseudonym N. Bourbaki was exposed as such in an article by an American mathematician, R. P. Boas. Bourbaki countered with an article claiming that R. P. Boas did not exist.

## JOHN VON NEUMANN (1903-1957)

One of the most dramatic advances in the twentieth century has been the development of computers. The digital computer was developed during World War II when mathematicians served the government in the war effort. Perhaps the greatest mathematician of the second quarter of the twentieth century was John von Neumann, a Hungarian. Von Neumann was responsible for the concept of a stored program digital computer. He, along with many other European scientists and mathematicians, emigrated to America in the 1930's.

Von Neumann wrote on set theory logic and the mathematical foundations of quantum mechanics. He also founded the subjects of mathematical economics and game theory, among his many other contributions. Von Neumann became interested in how the human brain functions, and tried to find patterns of connection for neurons which would be able to store and transmit information. His studies of the brain and of logic proved useful in his research leading to the development of the computer.

Studying the structure of the brain seems to be a natural step for mathematicians; a step on a trail which von Neumann blazed. After all, what is the number 2? We do not see it anywhere; it is a concept of the mind. How does our brain represent concepts? It has been shown by the Austrian mathematician **Kurt Gödel** (1906— ), also an emigrant to America, that formal axiomatic systems cannot in themselves describe number completely. Yet all of us clearly understand this concept. How does the brain function? Or better, what is its structure?

Mathematics is concerned with various abstract structures. In a way, all that we know about the external world is its structure. Impressions are not transmitted exactly to the brain, but rather the structure of these impressions is. The air vibrating in a violin produces a sound. The structure of this music is represented in our brain by electrical impulses traveling across neurons, a different medium, yet the structure is preserved. Perhaps this is an area in which mathematics will develop in the future.

In reviewing the history of mathematics, we can appreciate some of the great achievements. The decimal system is superb for computation, especially in comparison to other number systems. We take the simple rules of algebra for granted, yet it took thousands of years for the symbolism $ax^2 + bx + c = 0$ to evolve. This algebra was used to advance many studies which were almost impossibly difficult without it, such as calculus. The rules of calculus represent a great achievement. The rise of abstract mathematics concerned with general structures has given mathematics great freedom and provided unforeseen applications. Many areas of study, in particular all social and natural sciences, are finding mathematical models quite useful, as are business and many other fields. We have seen that some men solved mathematical problems for pure pleasure, while others were primarily interested in understanding the world and the ways of its Creator. Probably both impulses are present to some extent in all the great mathematicians.

The great mathematicians, such as Pythagoras, Euclid, Archimedes, Descartes, Newton, Gauss, Riemann, etc., have had a great effect on society. Our conceptions of ourselves, of what it is to be human, of the nature of the world, have been shaped in part by these brilliant men. The fact that there were men who believed that knowledge of the world was possible gave a direction to Western culture that other cultures did not have; not every culture believes seeking truth is a reasonable endeavor.

I hope that this survey of the history of mathematics has given you an awareness that mathematics is not an eternal unchanging system, but is constantly changing and developing, subject to internal and external forces. I have presented events and topics of special interest to me — I like the story, and I hope you do, too.

## Problems

1. Graph the following complex numbers.
   a. $2 + 4\sqrt{-1}$
   b. $-3\sqrt{-1}$
   c. $-3 + 2\sqrt{-1}$
   d. $-5 - 3\sqrt{-1}$

2. Find the product of each of the pairs of complex numbers below. Verify that the angle of the product is the sum of the angles of the factors and the length of the product is the product of the lengths of the factors.
   a. $1 + \sqrt{-1}$, $-2 + 2\sqrt{-1}$
   b. $1 + \sqrt{3}\sqrt{-1}$, $\sqrt{3} + \sqrt{-1}$
   c. $3\sqrt{-1}$, $-1 + \sqrt{-1}$

3. The length $\sqrt{a}$ can be constructed with a straightedge and compass (fig. 6.37). Let $AB = a$ and $BC = 1$. Bisect $AC$ at $O$. Construct a circle with center $O$ and radius $OC$. Construct $BD$ perpendicular to $AC$. Prove algebraically that $BD = \sqrt{a}$.

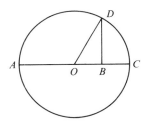

Figure 6.37

4. Gauss developed an ingenious method for solving the equations $z^p - 1 = 0$ where $p = 2^{(2^n)} + 1$ is prime. Let $n = 1$, so that $p = 5$. Let a regular pentagon be inscribed in a circle of radius 1 in the complex number plane (fig. 6.38).

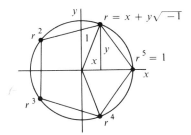

Figure 6.38

a. Use the properties of complex number multiplication to show that if $r$ is the vertex at an angle of 72°, then $r^2$ is the vertex at 144°, $r^3$ the one at 216°, and $r^4$ the one at 288°. Also show that all five vertices — 1, $r$, $r^2$, $r^3$, and $r^4$ — are roots of the equation $z^5 - 1 = 0$.

b. Gauss, starting with $r$, arranged the vertices, except the vertex 1, in a sequence in which each is the square of the preceding, i.e. — $r, r^2, r^4, r^8 = r^5 r^3 = r^3$. He then formed two sums each consisting of alternate terms of the sequence, i.e. — $A = r + r^4$ and $B = r^2 + r^3$. Show that $A + B = -1$ and $AB = -1$. [*Hint:* Factor $z^5 - 1 = 0$.]

c. Solve the equations in part b to show that $A = (-1 + \sqrt{5})/2$. Since $r = x + y\sqrt{-1}$, and $r^4 = x - y\sqrt{-1}$, then $A = 2x$, and $x = (-1 + \sqrt{5})/4$. Because this expression for $x$ contains only square roots, the length $x$ can be constructed with a straightedge and compass. The vertex of a regular pentagon is obtained by constructing a perpendicular from $x$ to the circle.

5. Gauss used an analogous procedure to that in problem 4 to show that the 17-sided polygon could be constructed with a straightedge and compass.

   a. Use the properties of complex number multiplication to show that if $r$ is the vertex at an angle of 360/17°, then the other vertices are given by $r^2, r^3, r^4, \ldots, r^{15}, r^{16}$, and $r^{17} = 1$. Also show that all 17 vertices — $r, r^2, r^3, \ldots, r^{16}$, and 1 — are roots of $z^{17} - 1 = 0$.

   b. Show that if you arrange the vertices, starting with $r$, in a sequence in which each is the square of the preceding, then the sequence starts to repeat itself after only eight vertices have been obtained. [*Note:* Remember to use the fact that $r^{17} = 1$ to simplify.]

   c. Show that if the vertices are arranged, starting with $r$, in a sequence in which each is the cube of the preceding, then all sixteen vertices unequal to 1 are obtained in the order $r, r^3, r^9, r^{10}, r^{13}, r^5, r^{15}, r^{11}, r^{16}, r^{14}, r^8, r^7, r^4, r^{12}, r^2, r^6$.

   d. Take alternate terms of the sequence in part c to obtain

   $$A_1 = r + r^9 + r^{13} + r^{15} + r^{16} + r^8 + r^4 + r^2$$

   and

   $$A_2 = r^3 + r^{10} + r^5 + r^{11} + r^{14} + r^7 + r^{12} + r^6$$

   Show that $A_1 + A_2 = -1$ and $A_1 A_2 = -4$, so $A_1$ and $A_2$ are the roots of $z^2 + z - 4 = 0$. It can be shown that $A_1$ is the positive root $1/2(\sqrt{17} - 1)$, and $A_2$ the negative root $1/2(-\sqrt{17} - 1)$.

e. Let $A_{11} = r + r^{13} + r^{16} + r^4$ and $A_{12} = r^9 + r^{15} + r^8 + r^2$. Show that $A_{11} + A_{12} = A_1$ and $A_{11}A_{12} = -1$, so $A_{11}$ and $A_{12}$ are roots of the equation $z^2 - A_1 z - 1 = 0$. It can be shown that $A_{11}$ is the positive root, $(1/2)A_1 + \sqrt{1 + (1/4)A_1^2}$.

f. Let $A_{21} = r^3 + r^5 + r^{14} + r^{12}$ and $A_{22} = r^{10} + r^{11} + r^7 + r^6$. Show that $A_{21} + A_{22} = A_2$ and $A_{21}A_{22} = -1$, so $A_{21}$ and $A_{22}$ are roots of the equation $z^2 - A_2 z - 1 = 0$. It can be shown that $A_{21}$ is the positive root, $(1/2)A_2 + \sqrt{1 + (1/4)A_2^2}$.

g. Let $A_{111} = r + r^{16}$ and $A_{112} = r^{13} + r^4$. Show that $A_{111} + A_{112} = A_{11}$ and $A_{111}A_{112} = A_{21}$, so that $A_{111}$ and $A_{112}$ are roots of the equation $z^2 - A_{11}z + A_{21} = 0$. It can be shown that $A_{111}$ is larger than $A_{112}$. A method for construction of the 17-sided polygon based on the analysis in this problem is given in problem 7.

6. Let a circle be constructed with diameter given by the line segment between the points $(0,1)$ and $(a,b)$ (fig. 6.39).

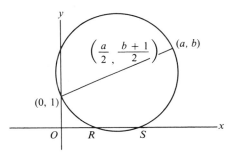

Figure 6.39

a. Show that the center of the circle is $(a/2, (b+1)/2)$.
b. Show that the equation of the circle is

$$\left(x - \frac{a}{2}\right)^2 + \left(y - \frac{b+1}{2}\right)^2 = \left(\frac{a}{2}\right)^2 + \left(\frac{b-1}{2}\right)^2$$

c. Show that the points of intersection, $R$ and $S$, of the circle with the line $y = 0$ satisfy the equation $x^2 - ax + b = 0$. Thus, the construction in this problem provides the solutions with a straightedge and compass of the equation $x^2 - ax + b = 0$.

7. A construction with a straightedge and compass of the 17-sided polygon can be devised based on the analysis of problem 5. Construct a circle of radius 1 with perpendicular diameters $AB$ and $DC$ (fig. 6.40). Construct the tangents to the circle at $D$ and $A$, which intersect at $S$. Construct $EA = (1/4)AS$ by bisecting $AS$ twice. With

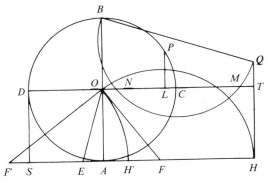

Figure 6.40

$E$ as center and $EO$ as radius construct a circle, and let $F$ and $F'$ be the points of intersection of the circle and $AS$. Let the circle with $F$ as center and radius $FO$ cut $AS$ at $H$, and let the circle with center $F'$ and radius $F'O$ cut $AS$ at $H'$. Construct $HTQ$ parallel to $AB$, where $T$ is its point of intersection with $OC$ (extended). Let $TQ = AH'$. Construct the circle with diameter $BQ$, and call $M$ and $N$ its points of intersection with $OC$, where $OM$ is greater than $ON$. Bisect $OM$ at $L$. Construct a perpendicular to $L$, intersecting the circle at $P$. The segment $PC$ is the side of a regular 17-sided polygon.

In parts $a$, $b$, and $c$, $A_1$, $A_2$, $A_{11}$, $A_{21}$, $A_{111}$, and $A_{112}$ are defined as in problem 5.

a. Show that

$$OE = \frac{1}{4}\sqrt{17} \qquad OF = \sqrt{1 + \frac{1}{4}A_1^2}$$

$$AF = \frac{1}{2}A_1 \qquad OF' = \sqrt{1 + \frac{1}{4}A_2^2}$$

$$AF' = -\frac{1}{2}A_2$$

b. Show that $AH = A_{11}$ and $AH' = A_{21}$.

c. Use problem 6 to show that $ON$ and $OM$ are the roots of the equation

$$z^2 - A_{11}z + A_{21} = 0$$

Thus, by problem 5g, the larger root $OM$ equals $A_{111}$, so that $OM = r + r^{16}$. If $r = x + y\sqrt{-1}$, then $r^{16} = x - y\sqrt{-1}$, so $OM = 2x$. Thus, $x = OM/2$ which equals $OL$, so that $P$ is, in fact, a vertex of the 17-sided polygon.

8. Show that a 90° angle can be trisected with a straightedge and compass.

9. If $x^3 - 2$ could be factored into the product of two polynomials with rational number coefficients, then one factor would have to be of degree one, say $bx - a$. Then the equation $x^3 - 2 = 0$ would have the root $x = a/b$. Show that $a/b$ is not a root of $x^3 - 2 = 0$, thus also proving that $x^3 - 2$ is irreducible. [*Hint:* Suppose $(a/b)^3 - 2 = 0$ and obtain a contradiction, as in the proof of the impossibility of $(a/b)^2 = 2$.]

10. Show that $x^3 - 3x - 1 = 0$ has no rational root $x = a/b$, thus $x^3 - 3x - 1$ is irreducible. [*Hint:* Assume $a/b$ is a root and show that $a^3/b - 3ab - b^2 = 0$. From that show that since $a$ and $b$ are assumed to have no factors except unity in common, $b = \pm 1$. Then show that if $a$ is an integer, then $a = \pm 1$. Thus, $x$ must equal $\pm 1$, but neither value gives a solution.]

11. Prove that if $a \equiv b \pmod{m}$ and $c \equiv d \pmod{m}$, then $a + c \equiv b + d \pmod{m}$. [*Hint:* By the definition of congruence, $a - b = mn_1$.]

12. a. Use Fermat's theorem to show that 47 is the first prime which can possibly divide $2^{23} - 1$.
    b. Use congruences to show that 47 does divide $2^{23} - 1$, so that $2^{22}(2^{23} - 1)$ is not a perfect number.

13. a. Use Fermat's theorem to show that 103, 137, 239, and 307 are the only primes smaller than $\sqrt{2^{17} - 1}$ which could possibly divide $2^{17} - 1$.
    b. Use congruences to show that none of the primes in part a divide $2^{17} - 1$. Thus, $2^{17} - 1$ is prime, and $2^{16}(2^{17} - 1)$ is a perfect number.

14. a. Use Fermat's theorem to show that 191, 229, 419, 457, 571, and 647 are the only primes smaller than $\sqrt{2^{19} - 1}$ which could possibly divide $2^{19} - 1$.
    b. Use congruences to show that none of the primes in part a divide $2^{19} - 1$. Thus, $2^{19} - 1$ is prime, and $2^{18}(2^{19} - 1)$ is a perfect number.

    (This result and that in problem 13 were first found by **Pietro Cataldi** in 1607. Of course, he did not use congruences. The next perfect number, $2^{30}(2^{31} - 1)$, was found by Euler, who showed that $2^{31} - 1 = 2,147,483,647$ is prime. After that, the next perfect number is $2^{60}(2^{61} - 1)$. The number $2^{61} - 1 = 2,305,843,009,-213,693,951$ was shown to be prime toward the end of the nineteenth century. Using computers, numbers such as $2^{19,937} - 1$, which has 6002 digits, have been shown to be prime.)

15. a. Use Fermat's theorem to show that 59 and 233 are the smallest two primes which could possibly divide $2^{29} - 1$.
    b. Use congruences to show that $2^{29} - 1$ is divisible by 233, so that $2^{29} - 1$ is not prime.

16. Compute the periods of the decimals for the following fractions, verifying each time that the period length of the fraction $1/m$ divides $m - 1$.
    a. $1/11$     b. $1/17$     c. $1/23$

17. Using congruences, obtain the periods for the fractions in problem 16 by finding the smallest positive divisor $p$ of $m - 1$, such that $10^p \equiv 1 \pmod{m}$.

18. Let any triangle be inscribed in a conic section. Let each side be extended to intersect the tangent to the conic at the opposite vertex. Prove that these three points of intersection lie in a straight line. [*Hint:* This is a special case of Pascal's theorem. As the points $B$ and $C$ of the hexagon get closer together, the segment $BC$ approaches a line tangent to the conic. Let the points $B$ and $C$, $D$ and $E$, and $A$ and $F$ approach one another, and the hexagon becomes a triangle, and the three segments become tangents.]

19. In projective geometry the dual of the statement, "a point lies on a conic," is the statement, "a line lies on a conic," meaning that the line is tangent to the conic. By interchanging point and line to form the dual of Pascal's theorem, the six vertices becomes six tangents. The lines connecting the vertices become the points of intersection of the tangents. In this process what do the points of intersection of the opposite sides become? State the dual of Pascal's theorem.

20. The Fourier series for the function shown in figure 6.41 is

$$f(x) = \sin x + \frac{\sin 3x}{3} + \frac{\sin 5x}{5} + \frac{\sin 7x}{7} + \cdots$$

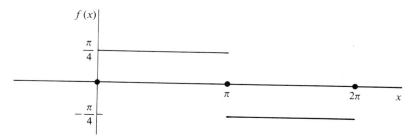

Figure 6.41

a. As a crude first approximation to the graph, sketch the curve $y = \sin x$.
b. As a slightly better approximation, sketch the graph of $y = \sin x + (\sin 3x)/3$. [*Hint:* First sketch $y = \sin x$, then on the same graph sketch $y = (\sin 3x)/3$ and add the two graphs together.]
c. As an even closer approximation, sketch the graph of $y = \sin x + (\sin 3x)/3 + (\sin 5x)/5$. [*Note:* Each of these approximations is continuous, in contrast to the infinite series $f(x)$ which has a jump at $x = \pi$.]

21. Compute the following products of complex numbers using Hamilton's definition.
    a. $(3,2)(0,2)$
    b. $(-3,4)(-3,-4)$
    c. $(0,-3)(0,-3)$
    d. $(3,2)(4,3)$

22. Quaternions can be defined using three quantities — $i, j, k$ — which satisfy the relations

$$jk = i \qquad kj = -i \qquad ki = j \qquad ik = -j$$
$$ij = k \qquad ji = -k \qquad i^2 = j^2 = k^2 = -1$$

Multiply the following quaternions using these relations.
a. $(2 + 3i + 4j + 3k)(1 + i + 2j + 3k)$
b. $(3 - 2i - j + k)(3 - i + 2k)$   c. $(2i - 3j + k)(1 - 2k)$

23. Find the following matrix products.

a. $\begin{bmatrix} 3 & 4 \\ 2 & 7 \end{bmatrix} \begin{bmatrix} -1 & 2 \\ 3 & -2 \end{bmatrix}$   b. $\begin{bmatrix} 4 & -1 \\ -2 & 0 \end{bmatrix} \begin{bmatrix} 3 & 2 \\ 0 & 6 \end{bmatrix}$

c. $\begin{bmatrix} 3 & 4 & -1 \\ 0 & 2 & 0 \\ 3 & 1 & 0 \end{bmatrix} \begin{bmatrix} 4 & 6 & 3 \\ 2 & 0 & 3 \\ -1 & 0 & -1 \end{bmatrix}$   [*Hint:* Use a rule analogous to that for $2 \times 2$ matrices.]

24. Let the sets $A$, $B$, and $C$ be represented by circles (fig. 6.42).
a. Shade in the area in figure 6.42 which represents $B \cap C$. Then shade in $A$. The total shaded region represents $A \cup (B \cap C)$.
b. On another diagram like figure 6.42, shade in $(A \cup B) \cap (A \cup C)$ confirming that it is equal to the set in part a.

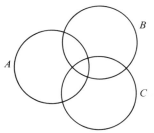

Figure 6.42

**25. a.** Show that the total length of the intervals removed in constructing the Cantor set is

$$\frac{1}{3} + 2\left(\frac{1}{3}\right)^2 + 4\left(\frac{1}{3}\right)^3 + 8\left(\frac{1}{3}\right)^4 + \cdots$$

**b.** Use the formula $1 + r + r^2 + r^3 + \cdots = 1/(1-r)$ to show that the sum of the series in part a is 1, so that the Cantor set, which is what is left after the removal of these intervals, has length zero.

**26.** In our decimal system, numbers are represented using powers of ten. Thus, $364 = 3 \times 100 + 6 \times 10 + 4$. In the binary system, numbers are represented using powers of 2. Thus, 13 in the decimal system would be $(1 \times 8) + (1 \times 4) + (0 \times 2) + (1 \times 1)$, or 1101, in the binary system. Write the following decimal numbers in the binary system.

a. 7  b. 9  c. 15  d. 35
e. 64  f. 43  g. 108

**27.** The operations of addition, subtraction, multiplication, and division can be carried out in the binary system. Since all numbers are written using only 0s and 1s (see problem 26) these operations are very easy. For example,

```
    1010          110           11
  + 1101        × 11       11)1001
    10111         110          11
                  110          11
                  10010        11
                               11
```

Perform the following calculations in the binary system.
a. $11011 + 11011$  b. $1010 \times 101$
c. $111101 \div 101$  d. $1101101 - 1001$

256 Mathematics as Free Creation

**28.** In the decimal system fractions can be expressed using descending powers of 10. Thus, $1/3 = 3/10 + 3/100 + 3/1000 + \cdots$, or $.33333\ldots$. In the binary system fractions can be expressed using descending powers of 2. These expressions can be found by long division, just as $1/3 = .333\ldots$ is found by dividing $3\overline{)1.000\ldots}$. Express the following fractions using descending power of 2.

a. $\dfrac{2}{3}$ [*Hint:* Find $11\overline{)10.000\ldots}$ in binary.]   b. $\dfrac{3}{5}$

c. $\dfrac{2}{7}$   d. $\dfrac{4}{9}$

**29.** Not only fractions (see problem 28), but any real number, can be expressed in the binary system. For example, any point $b$ on the line between 0 and 1 has an expansion $b = .b_1 b_2 b_3 b_4 \cdots = b_1/2 + b_2/4 + b_3/8 + b_4/16 + \cdots$. For the $b$ shown in figure 6.43,

Figure 6.43

$b_1 = 0$ since $b$ is in the left half when the segment $[0, 1]$ is divided in half.

$b_2 = 1$ since $b$ is in the right half when the segment $[0, 1/2]$ is divided in half.

$b_3 = 0$ since $b$ is in the left half when $[1/4, 1/2]$ is divided in half.

Use the fact that every point in the line segment $[0, 1]$ has a binary expansion to show that there are more such points than there are positive integers. [*Hint:* Suppose there are just as many points as integers. Then the sets could be corresponded in some way, say,

| Integers | Points |
|---|---|
| 1 | .101110 ... |
| 2 | .010110 ... |
| 3 | .0010011 ... |
| . | . |
| . | . |
| . | . |

Show that any such correspondence is impossible by specifying a way of determining for any given list a point not on that list.]

**30. a.** Show that every real number of the form

$$\frac{a_1}{3} + \frac{a_2}{3^2} + \frac{a_3}{3^3} + \cdots$$

where each $a_n = 0$ or 2 is not in one of the intervals removed in constructing the Cantor set, and is, therefore, in the Cantor set.

**b.** Show that there are at least as many points in the Cantor set as there are in the whole interval [0, 1]. [*Hint:* As in problem 29, every real number in [0, 1] can be written in the form $b_1/2 + b_2/2^2 + b_3/2^3 + \cdots$ where each $b$ is either 0 or 1. Correspond to each such number an element of the Cantor set by letting $a_n = 2b_n$. Note that some numbers have two binary representations, just as $1/10 = .10000\ldots = .099999\ldots$ in the decimal system. For example, $1/2 = .10000\ldots = .01111\ldots$. To represent each number uniquely in binary, exclude one or the other of the two possibilities, say those which have all ones beyond some point.]

**31.** Give examples of sets with the following order types.
  **a.** $\omega + 2$  **b.** $2 + \omega$  **c.** $\omega + \omega + \omega$

**32.** Show that $2 + \omega = \omega$, but $\omega + 2 \neq \omega$.

**33.** Construct three maps, each of which requires four colors to draw.

**34.** Let $S$ be any nonempty set. Show that the function $d$ given by

$$d(a,b) = 0 \quad \text{if } a = b$$

and

$$d(a,b) = 1 \quad \text{if } a \neq b$$

satisfies the four properties required of a metric space. This is a trivial example in which the distance between any two unequal objects is one.

**35.** Find

$$d(f,g) = \int_0^1 |f(x) - g(x)| \, dx$$

where $f(x) = x$ and $g(x) = x^2$.

**36.** Show that the first four properties of a ring are satisfied by the set of $2 \times 2$ matrices in example 3 on page 246. [*Hint:* Assume that the corresponding properties are satisfied for the set of numbers.]

# References

### A Forerunner — Carl Friedrich Gauss (1777–1855)

Bell
Dickson
Dunnington
Gauss (1,2,3)
Hall

Kazarinoff
Lanczos
Shanks
Smith (2)
Struik (3)

### Advanced Calculus

Bell
Fourier
Grattan-Guinness

Manheim
Ore
Struik (3)

### Variety in Geometry

Bell
Bonola
Coolidge
DeLong

Kline (2)
Lanczos
Lobachevsky
Smith (2)

### Variety in Algebra

Bell
Boole
Boyer (1)
Dehn

Hamilton
Kramer
Smith (2)

### Arithmetization of Analysis

Bell
Cantor
Dedekind
Eves (1)
Grattan-Guinness

Hawkins
Kramer
Manheim
Smith (2)

### Generality of Mathematics in the Twentieth Century

Bochner
Kramer
Hilbert
Manheim

Reid
Rosenblueth
von Neumann

# Appendices

# A  Suggestions For Further Reading

Works in this section are listed by author; refer to the Bibliography for titles and publishers. Note that references to specific periods and topics are listed at the end of each chapter in the text.

1. History of topics taught in the schools
   a. National Council of Teachers of Mathematics' *Historical Topics for the Mathematics Classroom* contains suggestions for the use of history in the classroom. Other chapters treat numerals, computation, algebra, geometry, trigonometry, and calculus.
   b. Smith (1), Vol. II contains a topic-by-topic discussion of school mathematics.
   c. Read provides a bibliography of articles on various topics in the history of mathematics. The *Cumulative Index: The Mathematics Teacher* contains a listing of articles on the history of mathematics which have appeared in the *Mathematics Teacher*.
2. Biographies
   a. Bell has written lively biographies of mathematicians from the period 1600–1900.
   b. The *Dictionary of Scientific Biography* contains articles on the life and work of mathematicians and scientists.
   c. Dunnington, Hall, Ore, and Reid have all written biographies of mathematicians.
   d. The bibliographies in topics 1c and 5 have many references to biographical articles.
3. History of Mathematical Developments
   a. Aaboe has excellent chapters on several areas of Greek and Babylonian mathematics.
   b. Van der Waerden writes a thorough discussion of the development of Greek mathematics, along with chapters on the Egyptians and the Babylonians.
   c. Kline (1) is an advanced work which extensively treats mathematics from 1700 to the present.
   d. Kramer presents an introductory treatment of topics in modern mathematics with much historical background.

4. Primary Sources
    a. Euclid is the most accessible and is quite fascinating. There is no better way to understand the history of mathematics than to read the original works. A number of other complete works are available and require varying degrees of mathematical background.
    b. Smith (2), Struik (3), Midonick, Thomas, and Wolff provide collections of excerpts from original works. Wolff includes commentaries on each selection and could be read in conjunction with this book.
5. General Bibliography
    May provides a complete bibliography for the history of mathematics. His purpose is to assist mathematicians, users of mathematics, and historians in finding and communicating information required for research, applications, teaching, and exposition.

# B  Suggestions For Projects

The following list contains suggestions for extended projects. Some suggested projects refer to a specific historical period, while others deal with one topic over an extended time. The list is by no means exhaustive.

1. Referring to Euclid's *Elements*, explain how he constructed the various regular polygons.
2. Give the history of some computing devices, such as the abacus, slide rule, adding machines, computers, etc.
3. Give the history of the three famous problems of Greek geometry and of various solutions to them.
4. Give the history of quadratic equations and their solutions.
5. Report on the history of the number $\pi$, including the approximate values for $\pi$ used by various peoples.
6. Give a history of the calendar.
7. Describe Egyptian results in geometry.
8. Describe Babylonian results in geometry.
9. Explore early Greek philosophy and its relationship to mathematical concepts.
10. Report on the life of a mathematician, such as Gauss, Abel, Cardano, Newton, Descartes, Hilbert, etc.
11. Give a history of the notation and symbols of algebra.
12. Give a history of perfect numbers.
13. Explore the development of non-Euclidean geometry.
14. Give a history of the different types of number systems that have been used throughout history.
15. Give examples of how the history of mathematics might be used in the classroom.
16. Compare the development of mathematics with that of art. (*See* Ivins.)
17. Give a history of perspective painting.
18. Make models of important mathematical objects. For example, if a track is made in the shape of a cycloid, the tautochrone property

could be demonstrated by letting two ball bearings roll from different heights on either side of the track. Both balls will reach the bottom at the same time.

19. Report on a portion of one of the works in topic 3 or 4 of the Suggestions for Further Reading.
20. Report on several articles chosen using the bibliographies named in topic 1c of the Suggestions for Further Reading.
21. Compute the complete table of chords using Ptolemy's methods. If you use a circle of radius one and the decimal system, then you can check your results by referring to a table of sines. Use a computer, unless you truly like calculating.
22. Learn how the Fibonacci numbers were used by Lucas in his method for finding if large numbers are prime. (*See* Shanks.)

# C Guide to the Pronunciation of Names

Explanations of the symbols used can be found in *Webster's Biographical Dictionary*, *Webster's New International Dictionary*, or *Webster's New Collegiate Dictionary*.

## Greek

| | | | |
|---|---|---|---|
| Apollonius | ăp′ ŏ·lō′ nĭ·ŭs | Heron | hēr′ ŏn |
| Archimedes | är′ kĭ·mē′ dēz | Hipparchus | hĭ·pär′ kŭs |
| Archytas | är·kī′ tăs | Hippocrates | hĭ·pŏk′ rȧ·tēz |
| Aristotle | ăr′ ĭs·tŏt′ 'l | Hypatia | hī·pā′ shĭ·ȧ |
| Boethius | bō·ē′ thĭ-ŭs | Nicomachus | nĭ·kŏm′ ȧ·kŭs |
| Democritus | dē·mŏk′ rĭ·tŭs | Plato | plā′ tō |
| Diophantus | dī′ ō·făn′ tŭs | Proclus | prō′ klŭs |
| Eratosthenes | ĕr′ ȧ·tŏs′ thē·nēz | Ptolemy | tŏl′ ĕ·mĭ |
| Euclid | ū′ klĭd | Pythagoras | pĭ·thăg′ ō·răs |
| Eudemus | û·dē′ mŭs | Thales | thā′ lēz |
| Eudoxus | û·dŏk′ sŭs | Zeno | zē′ nō |

## Middle Eastern and Indian

| | | | |
|---|---|---|---|
| Abu'l Wefa | ȧ·bōōl′ wĕ·fä′ | Aryabhata | är′ yȧ·bŭt′ ȧ |
| Ahmes | ä′ mĕs | Bhaskara | bäs′ kȧ·rȧ |
| Al-Battani | ăl′ băt·tä′ nē | Brahmagupta | brŭ′ mȧ·gōōp′ tȧ |
| Al-Khowarizmi | ăl·kōō·wä′ rĭz·mē | Omar Khayyám | ō′ mär kī·(y)äm′ |

## European

| | | | |
|---|---|---|---|
| Abel | ä′ bĕl | Cavalieri | kä′ vä·lyâ′ rē̆ |
| Adelard | ăd′ ĕ·lärd | Ceulen | kû′ lĕn |
| Argand | ȧr′ gäN′ | Chuquet | shü′ kĕ′ |
| Bede | bēd | Copernicus | kō·pûr′ nĭ· kŭs |
| Bernoulli | bĕr·nōōl′ ē̆ | Crelle | krĕl′ ĕ |
| Bolzano | bōl·tsä′ nō | D'Alembert | dȧ′ läNbâr′ |
| Bradwardine | brăd′ wĕr·dēn | Dedekind | dā′ dĕ·kĭnt |
| Brianchon | brē′ äN′ shôN′ | De Moivre | dē·mwȧ′ vr′ |
| Cardano | kär·dä′ nō | Desargues | dā′ zȧrg′ |
| Cauchy | kō′ shē′ | Descartes | dā·kärt′ |

| | | | |
|---|---|---|---|
| Dürer | dü′ rėr | Monge | môNzh |
| Euler | oi′ lēr | Napier | nā′ pĭ·ėr |
| Fermat | fėr′ mȧ′ | Oresme | ô′ râm′ |
| Fibonacci | fē′ bō·nät′ chē | Oughtred | ô′ trĕd |
| Fiore | fyō′ rā̂ | Pacioli | pä·chō′ lē̂ |
| Fourier | fo͞o′ ryā′ | Pascal | pȧs′ kȧl′ |
| Francesca | frän·chäs′ kä | Planudes | plɑ·nū′ dēz |
| Galileo | gä′ lē̂·lä′ ō̂ | Poncelet | pôNs′ lē̂′ |
| Galois | gȧ′ lwȧ′ | Recorde | rĕk′ ôrd |
| Gauss | gous | Regiomontanus | rē′ jĭ·ō̂·mŏn·tā′ nŭs |
| Gerard | jėr·ȧrd′ | | |
| Gerbert | zhĕr′ bâr′ | Riemann | rē′ män |
| Hadamard | ȧ′ dȧ′ mȧr′ | Riese | rē′ zĕ |
| Hermite | ėr′ mēt′ | Roberval | rô′ bĕr ′vȧl′ |
| Huygens | hoi′ gĕns | Sacrobosco | săk′ rō̂·bŏs′ kō |
| Kummer | ko͝om′ ėr | Stevin | stĕ·vīn′ |
| Lagrange | lȧ′ gräNzh′ | Tartaglia | tär-tä′ lyä |
| Laplace | lȧ′ plȧs′ | Torricelli | tōr′ rē̂·chĕl′ lē̂ |
| Leibniz | līp′ nĭts | Viète | vyĕt |
| L'Hospital | lô′ pē′tȧl′ | Von Neumann | fôn noi′ män |
| Lindemann | lĭn′ dĕ·män | Weierstrass | vī′ ėr·shträs′ |
| Lobachevsky | lō′bɑ·chĕf′ skĭ | Wessel | vĕs′ ĕl |
| Mercator | mûr·kā′ tēr | Widman | vĭt′ män |
| Möbius | mû′ bē̂·o͞os | | |

# D  Answers to Selected Problems

## Chapter 1

1. 9 9 9 9 ∩ ∩ ∩ |
   9 9 9     ∩ ∩

2. a.  
   | √1 | 15 |
   | 2 | 30 |
   | 4 | 60 |
   | √8 | 120 |

   $15 + 20 = 135$

3. a. $\bar{2} + \bar{16}$;  b. $6 + \bar{4}$;  c. $\bar{2} + \bar{8} + \bar{16} + \bar{80}$  Other answers are possible.
4. $10 + \bar{3}$  5. $(32/3)^2$  6. $\bar{4} + \bar{28}$  7. $\bar{6} + \bar{18}$; $\bar{8} + \bar{18} + \bar{36} + \bar{72}$  Other solutions are possible. The only other with two terms is $\bar{5} + \bar{45}$.  8. a. $\bar{3} + \bar{30}$; b. $\bar{3} + \bar{10} + \bar{30}$; c. $\bar{3} + \bar{5} + \bar{30}$  9. 4
10. a. ∨ ⫷∨  c. ⫷⫷⫷ ∨∨∨∨ ⫷⫷⫷ ∨
11. a. ∨∨  (0;4)  c. ∨∨∨∨  ∨∨∨
12. 0;8,34  13. a. 1/2, 1/4, 1/5, 1/8;  b. 1/2, 1/3, 1/4, 1/5, 1/6, 1/8, 1/9
14. 1,2,23;3,15  15. 1,0;32,16  16. a. 2.65;  b. 3.46;  c. 4.47  17. 2
18. 0;45  19. 18 and 2  20. 7-24-25, 11-60-61, 9-40-41, 13-84-85  Many others are possible.  23. $\pi[(2 + \sqrt{2})/6]^2 s^2$, or approx. $1.02\, s^2$

## Chapter 2

1. 8128  2. 22
3. a. ⌐ΔΔ||  b. HHHH⌐ΔΔΔ⌐|||  c. M⌐HH⌐Δ⌐
The ⌐ represents 500.
4. a. $\overline{\kappa \varsigma}$;  b. $\overline{\phi \mu}$;  c. $,\gamma \upsilon \xi \theta$;  d. $M \cdot \eta \rho \pi \delta$  6. .62 and .38  7. a. 5;  b. 3;  c. 1
9. 2, 3, 5, 7, 11, 13, 17, 19, 23, 29, 31, 37, 41, 43, 47, 53, 59, 61, 67, 71, 73, 79, 83, 89, 97  10. $\sqrt{80}$  12. a. 2.44;  b. 2.64;  c. 4.47  13. a. $C^u 9\ U5\ LE\ S^q 3\ NU2$;  b. $NU3\ U2\ LE\ S^q 9$  24. b. $6/\sqrt{3} \approx 918/265$  25. $12/(2 + \sqrt{3}) \approx 1836/571$  28. $4\frac{4}{5}$  29. approx. 12.54  30. b. $EQ = 2;16$ and $FQ = 1;2$  34. b. $k = 3$ gives 6/5 and 17/5; $k = 4$ gives 29/17 and 54/17  35. There are many solutions, for example a. 4/5 and 22/5;  b. 7/5 and 24/5  36. b. $x = 20$, $z = 41$, $w = 80$, $y = 320$;  c. There are many, for example if $d = 7$, then $w = 32$, $y = 32$, and $z = 17$.  37. a. $1 = 2^0$, $2 = 2^1$, $4 = 2^2$, $8 = 2^3$, $16 = 2^4$, $31 = 2^0(2^5 - 1)$, $62 = 2^1(2^5 - 1)$, $124 = 2^2(2^5 - 1)$, $248 = 2^3(2^5 - 1)$, $496 = 2^4(2^5 - 1)$

## Chapter 3

2. 
| 1  | 15 | 14 | 4  |
|----|----|----|----|
| 12 | 6  | 7  | 9  |
| 8  | 10 | 11 | 5  |
| 13 | 3  | 2  | 16 |

Each sum is 34.

3. Here is a sequence of steps. Others are possible.

$$\begin{bmatrix} 1 & 0 & 3 \\ 2 & 5 & 2 \\ 3 & 1 & 1 \\ 26 & 24 & 39 \end{bmatrix} \to \begin{bmatrix} 3 & 0 & 3 \\ 6 & 5 & 2 \\ 9 & 1 & 1 \\ 78 & 24 & 39 \end{bmatrix} \to \begin{bmatrix} 0 & 0 & 3 \\ 4 & 5 & 2 \\ 8 & 1 & 1 \\ 39 & 24 & 39 \end{bmatrix} \to \begin{bmatrix} 0 & 0 & 3 \\ 20 & 5 & 2 \\ 40 & 1 & 1 \\ 195 & 24 & 39 \end{bmatrix}$$

$$\to \begin{bmatrix} 0 & 0 & 3 \\ 20 & 20 & 2 \\ 40 & 4 & 1 \\ 195 & 96 & 39 \end{bmatrix} \to \begin{bmatrix} 0 & 0 & 3 \\ 0 & 20 & 2 \\ 36 & 4 & 1 \\ 99 & 96 & 39 \end{bmatrix} \to \begin{bmatrix} 0 & 0 & 3 \\ 0 & 5 & 2 \\ 36 & 1 & 1 \\ 99 & 24 & 39 \end{bmatrix}$$

4. a. 3; b. 2; c. 3; d. 8; e. 6; f. 5

```
 a    b    b    a    c    c
 5    4    7         3    4
      a              b
      9              4
                     a
                     2
```

5. $x^8 + 8x^7y + 28x^6y^2 + 56x^5y^3 + 70x^4y^4 + 56x^3y^5 + 28x^2y^6 + 8xy^7 + y^8$

6. a. ||| ≣   b. ||≡ ||||   c. |||○ ⊤
   d. ||||○   e. || ≐ |||| ⊥   f. ||||○⊤ ≣

7. a. *ya v* 3 *ya* 5̇
      *ru* 7
   b. *ya v* 2
      *ya* 4 *ru* 3̇

9. a. 1 from 2 is 1; 9 from 10 is 1, 1 and 8 is 9; 2 from 5 is 3

10. a.
```
         3   2
       ┌───┬───┐
     2 │1 ╱│1 ╱│ 6
       │ ╱8│ ╱2│
       ├───┼───┤
     0 │1 ╱│0 ╱│ 4
       │ ╱2│ ╱8│
       └───┴───┘
         4   8
```

11a.  
```
     1
     3̸
   2̸9̸7  ) 23 r.17
   6̸3̸8̸
   2̸7̸7̸
     2̸
```

12. a. •   b. ::::   c. ≡   d. :::

16. a. $x = 5/3$; b. The new solution is 19/11.   17. 100   18. Either 16 or 48.
19. 72

**Chapter 4**

4. a.
$$10 - 2\overline{)652} \quad 60 + 10 + 9 + 2 = 81 \text{ r.}4$$
$$\begin{array}{r} 600 \\ \hline 52 \\ +120 \\ \hline 172 \\ 100 \\ \hline 72 \\ +20 \\ \hline 92 \\ 90 \\ \hline 2 \\ +18 \\ \hline 20 \\ 20 \\ \hline 0 \\ +4 \end{array}$$

c.
$$40 - 2\overline{)5721} \quad 100 + 40 + 10 = 150 \text{ r.}21$$
$$\begin{array}{r} 4000 \\ \hline 1721 \\ +\ 200 \\ \hline 1921 \\ 1600 \\ \hline 321 \\ +80 \\ \hline 401 \\ 400 \\ \hline 1 \\ +20 \\ \hline 21 \end{array}$$

6. a. 8 from 11 is 3; 5 from 12 is 7; 8 from 16 is 8; 2 from 3 is 1. Answer is 1873

7. 1, 1, 2, 3, 5, 8, 13, 21, 34, 55, 89, 144, 233, 377, 610, 987, 1597, 2584, 4181, 6765, 10946, 17711, 28657, 46368, 75025

11. a. 13) 147   (11 r.4
$$\begin{array}{r} 1 \\ \hline 4 \\ 3 \\ \hline 17 \\ 1 \\ \hline 7 \\ 3 \\ \hline 4 \end{array}$$

12. a.  437  441  461  961
         524   52    5

13. a.  51
        246
        4264
        18682
        3777
        33
        ̄14134
        252

14. a.  14    7    3    1
        29   58  116  232
        232 + 116 + 58 = 406

15. a. $.6_4^4$; b. $.7^{5 \cdot \bar{m}}$; c. $.3_1^2 \bar{p} \; .4^{3 \cdot \bar{m}}$   16. a. Units digit is 8. Since $5 \cdot 4 + 3 \cdot 2 = 26$, the tens digit is 6. Then $3 \cdot 5 = 15$, and $15 + 2 = 17$, so the answer is 1768.
18. a. $\sqrt[3]{\sqrt{26} + 5} - \sqrt[3]{\sqrt{26} - 5}$; b. $\sqrt[3]{4} - \sqrt[3]{2}$
19. a. $\sqrt[3]{20 + \sqrt{392}} + \sqrt[3]{20 - \sqrt{392}}$  b. $\sqrt[3]{4} + \sqrt[3]{2}$
20a. 3 ⓪ 6 ① 8 ② 3 ③ ; b. 24 ⓪ 1 ① 0 ② 4 ③ 9 ④ ;
c. 576 ⓪ 3 ① 8 ② 8 ③ 4 ④ 2 ⑤   21. 1000 and 1000.10001
22. g. .0004950 is both an upper and a lower bound for the difference of the logarithms.   23. a. $.4 < \log 3 < .5$; b. $.47 < \log 3 < .48$; c. $.477 < \log 3 < .478$
24. After four more steps, log 3.995 (approx. log 4.00) is found to be .6016. Continuing another three steps gives log 4.004 = .6026. The true value is between these two values.

## Chapter 5

9. a. at most one positive and one negative; b. at most two positive and two negative; c. at most three positive and two negative   14. cross-ratio is 9/8 in both cases   16. $A$ gets \$1.75, $B$ gets \$.25   18. $\pi \sim 3.1416208$, carrying the calculation out to seven places   19. $1 - x^2 + x^4 - x^6 \ldots$   20. $1 + x + x^2 + x^3 \ldots$   23. a. 8; b. 3; c. 2   26. $4 = 2 + 2, 6 = 3 + 3, 8 = 5 + 3, 10 = 5 + 5, 12 = 7 + 5, 14 = 7 + 7, 16 = 11 + 5$, etc. Various other solutions are possible.   27. a. $\{s_1, s_2\}$; b. $\{s_1, s_3, s_6\}$; c. $\{s_1\}$

## Chapter 6

2. b. $1 + \sqrt{3}\sqrt{-1}$ has angle 60° and length 2; $\sqrt{3} + \sqrt{-1}$ has angle 30° and length 2. The product, $4\sqrt{-1}$, has angle 90° and length 4.   13. b. For example, $2^{17} - 1 \equiv 55 \pmod{103}$, so 103 does not divide $2^{17} - 1$.   16. a. $1/11 = .\overline{09} \ldots$ The period, 2, divides 10; b. $1/17 = .\overline{0588235294117647} \ldots$ The period, 16, divides 16; c. $1/23 = .\overline{0434782608695652173913} \ldots$ The period, 22, divides 22.   21. a. $(-4,6)$; b. $(25,0)$; c. $(-9,0)$; d. $(6,17)$   22. a. $-18 + 11i + 2j + 11k$; b. $5 - 11i + 8k$; c. $2 + 8i + j + k$

23. a. $\begin{bmatrix} 9 & -2 \\ 19 & -10 \end{bmatrix}$; b. $\begin{bmatrix} 12 & 2 \\ -6 & -4 \end{bmatrix}$; c. $\begin{bmatrix} 21 & 18 & 22 \\ 4 & 0 & 6 \\ 14 & 18 & 12 \end{bmatrix}$

26. a. 111; b. 1001; c. 1111; d. 100011; e. 1000000; f. 101011; g. 1101100

27. a. 110110;  b. 110010;  c. 1100 r.1;  d. 1100100    28. a. $.\overline{10}\ldots$; b. $.\overline{1001}\ldots$; c. $.0\overline{10}\ldots$; d. $.0\overline{11100}\ldots$    31. Some examples are a. $1/2, 2/3, 3/4, 4/5, \ldots, 1,2$; b. $-1,0,\overset{2}{\overbrace{1/2, 2/3, 3/4, 4/5, \ldots}}\overset{\omega}{\overbrace{\phantom{xxxxxxxx}}}$ c. $1/2, 2/3, 3/4, \ldots, 1\frac{1}{2}, 1\frac{2}{3}, 1\frac{3}{4}, \ldots, 2\frac{1}{2}, 2\frac{2}{3}, 2\frac{3}{4}, \ldots$    33. The maps of figure 6.26 b and c, for example.    35. 1/6

# Bibliography

# Bibliography

Aaboe, Asger. *Episodes from the Early History of Mathematics.* New York: Random House, 1964.

Apollonius. *Treatise on Conic Sections.* Edited by T. L. Heath. New York: Barnes and Noble, 1961.

Archimedes. *The Works of Archimedes.* Edited by T. L. Heath. New York: Dover Publications, 1953.

Baron, Margaret E. *The Origins of the Infinitesimal Calculus.* Oxford: Pergamon Press, 1969.

Beckmann, Petr. *A History of* II *(Pi).* Boulder: The Golem Press, 1971.

Bell, E. T. *Men of Mathematics.* New York: Simon and Schuster, 1937.

Bochner, Salomon. *The Role of Mathematics in the Rise of Science.* Princeton, N. J.: Princeton University Press, 1966.

Bonola, Roberto. *Non-Euclidean Geometry.* New York: Dover Publications, 1955.

Boole, George. *An Investigation of the Laws of Thought.* New York: Dover Publications, 1951.

Boyer, Carl B.
    (1) 1968. *A History of Mathematics.* New York: John Wiley and Sons.
    (2) 1959. *The History of the Calculus and its Conceptual Development.* New York: Dover Publications.

Bruins, E. M. *La Géométrie Non-Euclidienne dans L'Antiquité.* Paris: Palais de la Découverte, 1968.

Burnet, John. *Early Greek Philosophy.* Cleveland: World Publishing Co., 1957.

Burtt, E. A. *The Metaphysical Foundations of Modern Science.* Garden City: Doubleday and Co., 1932.

Cantor, Georg. *Contributions to the Founding of the Theory of Transfinite Numbers.* New York: Dover Publications, 1955.

Cardano, Girolamo. *The Great Art.* Translated by T. Richard Witmer. Cambridge, Mass.: M. I. T. Press, 1968.

Cohen, I. Bernard. *The Birth of a New Physics.* Garden City: Doubleday and Co., 1960.

Colebrooke, Henry Thomas, Esq., trans. *Algebra with Arithmetic and Mensuration from the Sanscrit of Brahmegupta and Bhascara,* Dr. Martin Sandig, 1973.

Coolidge, Julian Lowell. *A History of Geometrical Methods.* New York: Dover Publications, 1963.

Dedekind, Richard. *Essays on the Theory of Numbers.* Translated by W. W. Beman. La Salle, Ill.: Open Court Publishing Co., 1948.

Dehn, Edgar. *Algebraic Equations.* New York: Dover Publications, 1960.

Descartes, René. *The Geometry of René Descartes.* Translated by D. E. Smith and Marcia L. Latham. New York: Dover Publications, 1954.

Dickson, L. E. "Constructions with Ruler and Compasses; Regular Polygons." In *Monographs on Topics of Modern Mathematics,* edited by J. W. A. Young. New York: Longmans, Green and Co., 1924.

*Dictionary of Scientific Biography.* Charles Coulston Gillespie, Editor-in-Chief. New York: Charles Scribner's Sons, 1970– .

Diophantus. *Diophantus of Alexandria.* Edited by T. L. Heath. New York: Dover Publications, 1964.

Dörrie, Heinrich. *100 Great Problems of Elementary Mathematics.* New York: Dover Publications, 1965.

Dunnington, Guy. *Carl Friedrich Gauss, Titan of Science.* New York: Hafner Publishing Co., 1955.

Euclid. *The Thirteen Books of Euclid's Elements.* Edited by T. L. Heath. New York: Dover Publications, 1956.

Eves, Howard
   (1) 1969. *In Mathematical Circles.* Boston: Prindle, Weber and Schmidt.
   (2) 1969. *An Introduction to the History of Mathematics.* 3rd ed. New York: Holt, Rinehart and Winston.

Fourier, Jean Baptiste Joseph. *The Analytical Theory of Heat.* New York: Dover Publications, 1955.

Gandz, Solomon
   (1) 1937. "The Origin and Development of Quadratic Equations in Babylonian, Greek, and East Arabic Algebra." *Osiris* 3: 405–557.
   (2) 1931. "The Origin of the Ghubar Numerals, or the Arabian Abacus and the Articuli." *Isis* 3: 393–424.

Gauss, C. F.
   (1) 1966. *Disquisitiones Arithmeticae.* English translation by A. A. Clarke. New Haven, Conn.: Yale University Press.
   (2) 1965. *General Investigations of Curved Surfaces.* Translated by Adam Hiltebeitel and James Morehead. New York: Raven Press.
   (3) 1963. *Theory of Motion of Heavenly Bodies.* Translated by Charles Henry Davis. New York: Dover Publications.

Gillings, Richard J. *Mathematics in the Time of the Pharaohs.* Cambridge, Mass.: M. I. T. Press, 1972.

Grant, Edward. "Nicole Oresme and his *De proportionibus proportionum.*" *Isis* 51 (1960): 293–314.

Grattan-Guinness, Ivor. *The Development of the Foundations of Mathematical Analysis from Euler to Riemann.* Cambridge, Mass.: M. I. T. Press, 1970.

Hall, Tord. *Carl Friedrich Gauss.* Cambridge, Mass.: M. I. T. Press, 1970.

Hamilton, Sir William Rowan. *The Mathematical Papers.* Cambridge, Eng.: Cambridge University Press, 1931.

Hanson, N. H. "The Mathematical Power of Epicylical Astronomy." *Isis* 51 (1960): 150–58.

Haskins, C. H. *The Rise of the Universities.* Ithaca: Cornell University Press, 1957.

Hawkins, Thomas. *Lebesque's Theory of Integration; Its Origins and Development.* Madison: University of Wisconsin Press, 1970.

Heath, Sir Thomas L. *Greek Mathematics.* New York: Dover Publications, 1963.

Hilbert, David. *The Foundations of Geometry.* La Salle, Ill.: Open Court Publishing Co., 1962.

Hofmann, Joseph E. *The History of Mathematics.* New York: Philosophical Library, 1957.

Ivins, William. *Art and Geometry.* New York: Dover Publications, 1964.

Karpinski, Louis. *History of Arithmetic.* New York: Russell and Russell, 1965.

Kazarinoff, Nicolas C. *Ruler and the Round.* Boston: Prindle, Weber and Schmidt, 1970.

Klein, Jacob. *Greek Mathematical Thought and the Origin of Algebra.* Cambridge, Mass.: M. I. T. Press, 1968.

Kline, Morris
- (1) 1972. *Mathematical Thought from Ancient to Modern Times.* New York: Oxford University Press.
- (2) 1962. *Mathematics: A Cultural Approach.* Reading, Mass.: Addison Wesley Publishing Company.

Koestler, Arthur. *The Sleepwalkers.* New York: Macmillan and Company, 1959.

Kramer, Edna. *The Nature and Growth of Modern Mathematics.* New York: Hawthorn Books, 1970.

Lanczos, Cornelius. *Space Through the Ages.* London and New York: Academic Press, 1970.

Lasserre, Francois, *The Birth of Mathematics in the Age of Plato.* Larchmont, N. Y.: American Research Council, 1964.

Lattin, Harriet P. "The Origin of our Present System of Notation According to the Theories of Nicholas Bubnov." *Isis* 19 (1933): 181–94.

Lobachevsky, Nicholas. *Geometrical Researches on the Theory of Parallels.* Translated by G. B. Halsted in *Non-Euclidean Geometry* by Roberto Bonola. New York: Dover Publications, 1955.

Manheim, Jerome H. *The Genesis of Point Set Topology.* Oxford: Pergamon Press, 1964.

May, Kenneth O. *Bibliography and Research Manual of the History of Mathematics*. Toronto and Buffalo: University of Toronto Press, 1973.

Menninger, Karl. *Number Words and Number Symbols*. Cambridge, Mass.: M. I. T. Press, 1969.

Midonick, Henrietta, ed. *The Treasury of Mathematics; a collection of source material*. New York: Philosophical Library, 1965.

More, Louis T. *Isaac Newton, a Biography*. New York: Dover Publications, 1962.

Napier, John. *The Construction of the Wonderful Canon of Logarithms*. London: Dawsons of Pall Mall, 1966.

National Council of Teachers of Mathematics. *Cumulative Index: The Mathematics Teacher*, 1908–1965. Washington, D. C.: NCTM, 1967.

National Council of Teachers of Mathematics. *Historical Topics for the Mathematics Classroom*. 31st Yearbook. Washington, D. C.: NCTM, 1969.

Needham, Joseph. *Science and Civilization in China*, vol. 3. Cambridge: Cambridge University Press, 1959.

Neugebauer, O.
  (1) 1969. *The Exact Sciences in Antiquity*. New York: Dover Publications.
  (2) 1948. "The Astronomical Origin of the Theory of Conic Sections." *Proceedings of the American Philosophical Society* 92: 136–38.

Newton, Sir Isaac. *Mathematical Works*. Edited by Derek T. Whiteside. New York: Johnson Reprint, 1964.

Ore, Oystein. *Niels Henrik Abel, mathematician extraordinary*. Minneapolis: University of Minnesota Press, 1957.

Osen, Lynn. *Women in Mathematics*. Cambridge, Mass. and London: M. I. T. Press, 1974.

Perelman, Chaim. *An Historical Introduction to Philosophical Thinking*. New York: Random House, 1965.

Price, Derek, John de Solla. *Science Since Babylon*. New Haven: Yale University Press, 1961.

Ptolemy. *The Almagest. Great Books of the Western World*, vol. 16. Encyclopaedia Britannica, 1952.

Read, Cecil B. "The History of Mathematics: A Bibliography of Articles in English Appearing in Seven Periodicals." *School Science and Mathematics* 64 (1966): 147–79.

Regiomontanus. *On Triangles*. Translated by Barnabas Hughes. Madison: University of Wisconsin Press, 1967.

Reid, Constance. *Hilbert*. New York and Berlin: Springer-Verlag, 1970.

Robinson, John Mansley. *An Introduction to Early Greek Philosophy*. Boston: Houghton Mifflin Co., 1968.

Rosenblueth, Arturo. *Mind and Brain*. Cambridge, Mass.: M. I. T. Press, 1970.

Russell, Bertrand. *Wisdom of the West*. New York: Fawcett Publications, 1964.

Sambursky, Samuel. *The Physical World of the Greeks.* Translated by Merton Dagut. London: Routledge and Paul, 1960.

Sarton, George. *Ancient Science and Modern Civilization.* Lincoln: University of Nebraska Press. 1954.

Seidenberg, A.
  (1) 1962. "The Ritual Origin of Counting." *Archive for the History of the Exact Sciences* 2: 1–40.
  (2) 1961/62. "The Ritual Origin of Geometry." *Archive for the History of the Exact Sciences* 1: 488–527.
  (3) 1959. "Peg and Cord in Ancient Greek Geometry." *Scripta Mathematica* 24: 107–122.

Shanks, Daniel. *Solved and Unsolved Problems in Number Theory.* Washington, D. C.: Spartan Books, 1962.

Smith, David Eugene
  (1) 1958. *History of Mathematics.* 2 vols. New York: Dover Publications.
  (2) 1959. *A Source Book in Mathematics.* 2 vols. New York: Dover Publications.

Struik, D. J.
  (1) 1963. "Chinese Mathematics." *Mathematics Teacher* 56: 424–32.
  (2) 1958. "Omar Khayyam." *Mathematics Teacher* 51: 280–85.
  (3) 1969. *A Source Book in Mathematics, 1200–1800.* Cambridge: Harvard University Press.

Szabó, Árpád
  (1) 1964. "The Transformation of Mathematics into Deductive Science and the Beginning of its Foundation on Definitions and Axioms." *Scripta Mathematica* 27: 27–48, 113–39.
  (2) 1960. "Anfänge der Euklidishen Axiom." *Archive for the History of the Exact Sciences* 1: 37–106.

Thomas, Ivor. trans. *Greek Mathematical Works.* Cambridge: Harvard University Press, 1939.

Toeplitz, Otto. *The Calculus, A Genetic Approach.* Chicago: University of Chicago Press, 1963.

van der Waerden, B. L. *Science Awakening.* New York: Oxford University Press, 1963.

von Neumann, John. *The Computer and the Brain.* New Haven: Yale University Press, 1959.

Vorobyov, N. N. *The Fibonacci Numbers.* Boston: D. C. Heath and Co., 1963.

Wilder, R. L. *The Evolution of Mathematical Concepts.* New York: John Wiley and Sons, 1968.

Willerding, Margaret. *Mathematical Concepts: A Historical Approach.* Boston: Prindle, Weber and Schmidt, 1967.

Wolff, Peter. *Breakthroughs in Mathematics.* New York: New American Library, 1963.

# Index

# Index

Abacus, 97–100
Abel, Niels Henrik, 222, 231
Abu'l-Wefa, 115
Achilles paradox, 35
Adelard of Bath, 123
Addition
　abacus, 99
　Egyptian, 3
　Sacrobosco's method, 127
Adding machine, 174
Ahmes papyrus, 2, 7, 8
Al-Battani, 115
Alberti, Leon Battista, 135
Algebra
　abstract, 227–32, 245–47
　analytic art, 140
　Boolean, 232
　completing the square, 13, 112
　of Diophantus, 83
　double false position rule, 113, 117
　Egyptian, 8–9
　equations. *See* Equations
　fanciful problems, 117–18
　false position rule, 8–9
　group. *See* Group
　laws, 227–28, 245–46
　matrix, 229–31, 246
　noncommutative, 212, 229–31
　notation, 131–35
　permutations. *See* Permutations
　quaternions, 229, 255
　rings. *See* Ring
　set, 232
　symbolism, 83, 131–35, 140
　as symbol manipulation, 140
　word origin, 111
Algorithm, word origin, 111

Al-Kashi, 133
Al-Khowarizmi, 77, 111–14, 124
*Almagest*, 77
Analysis
　in algebra, 112, 140
　arithmetization of, 233–43
　in geometry, 139, 164–68
Analytic geometry, 139, 164–68, 171
Angle trisection. *See* Constructions
Apollonius, 73–75, 167, 174
Arabic mathematics, 111–15
Archimedes, 67–73, 76
Archytas, 45, 65
Arctangent, 176, 200
Area
　by calculus, 152–53, 168, 181
　circle, 8, 50–51
　isosceles triangle, 8
　octagon, 8
　parabolic segment, 72
　pentagon, 77, 91
　triangle, 76
Argand, Jean, 207
Aristotle, 53–54, 129
Arithmetic. *See* Addition, Subtraction, Multiplication, Division, and Square Root
Arithmetic terms, word origins, 134
Art, 135–36
Associative law, 228
Astrology, 81, 130, 131
Astronomy, 78, 90–91, 131, 135, 150, 178, 206
Axioms
　axiomatic systems, 243–47
　common notions, 37

Axioms, (*continued*)
   laws of algebra, 227–28, 245–46
   origin, 36–37

Babylonian mathematics, 9–17
Barrow, Isaac, 176
Bede, 122
Berkeley, George Bishop, 180
Bernoulli, Daniel, 186, 217
Bernoulli, James, 185–87
Bernoulli, John, 185–87, 188
Bhaskara, 106–9, 117
Binary numbers, 256–58
Binomial coefficients, 104, 174
Binomial theorem, 181
Bisection. *See* Constructions.
Boethius, 85, 123
Bolzano, Bernard, 222, 234
Bombelli, Rafael, 139
Boole, George, 232
Boolean algebra, 232
Bourbaki, Nicolas, 247
Bradwardine, Thomas, 129
Brachistochrone, 186
Brahmagupta, 106
Brain, 247
Brianchon, Charles-Julien, 215
Briggs, Henry, 148
Bubnov, Nicholas, 114

Calculating machine, 174
Calculus
   arc length, 182
   area, 168, 181–82
   center of gravity, 141
   completeness, 238–40
   differential, 168, 179–80
   integral, 152–53, 159, 168, 182–83, 221, 236
   limit, 50, 181, 185, 221, 222, 233
   notation, 184, 188
   precursors, 141, 152–53, 176
   word origin, 99
Cantor, Georg, 241–43
Cantor set, 242, 256, 258
Cardano, Girolamo, 137–38, 171, 222
Carroll, Lewis, 240
Cattle problem, 72

Cauchy, Augustin-Louis, 221–22, 236
Cavalieri, Bonaventura, 152–53, 159
Cayley, Arthur, 229–31
Celestial mechanics, 193
Center of gravity, 139, 141
Ceres, 214
Chinese mathematics, 101–4
Chord
   difference formula, 80
   sum formula, 91
   table, 75, 79–81
Chuquet, Nicolas, 132
Circumference of the earth, 66
Clocks, 176
Commensurability, 31–35
Common notions. *See* Axioms
Commandino, Federigo, 139
Commutative law, 227–31
Completeness, 238–40
Completing the square, 13, 112
Complex numbers. *See* Numbers, complex
Computers, 247
Cone, volume of, 52
Congruence, 212–13, 246
Conic sections, 53, 73–75, 173–74
   names of, 74
   solutions to four-line locus, 168
Constructions
   altars, 18, 19, 41
   bisection of a segment, 40
   doubling the square, 27
   duplication of the cube, 41, 45, 210
   product, 195
   quotient, 195
   rectangle equal to square, 19
   regular pentagon, 87, 249–50
   regular polygons, 42, 87, 208–10, 249–52
   right angle, 18
   17-sided polygon, 250–52
   solution to quadratic equations, 61–62, 195–96, 251
   square root, 249
   squaring the circle, 19, 41, 211
   trisection of an angle, 41, 69–70, 210, 253
Continued fractions, 125

Continuity, 185, 220, 221, 222, 234, 238–40
Convergence, 220, 222
Copernicus, Nicholas, 135
Cramer, Gabriel, 186
Crelle, August, 222
Cross-ratio, 172
Cuneiform, 10
Cycloid, 175–76, 186, 199

D'Alembert, Jean Le Rond, 217
De Moivre, Abraham, 186
De Morgan, Augustus, 228, 232
Decimals, 140
   decimal point, 133
   periodic, 212–13
Dedekind, Richard, 237–40
Degree, word origin, 134
Democritus, 53
Denominator, word origin, 134
Desargues, Girard, 171–73
Desargues' theorem, 172
Descartes, René, 163–68
Descartes' rule of signs, 197
Descriptive geometry, 194
Differential geometry, 214–15
Diophantus, 81–83, 92, 170
Dirichlet, Peter Gustav Lejeune, 223
Distance, 244–45
Distributive law, 228
Division
   abacus, 123
   a danda method, 133
   Egyptian, 5–7
   galley method, 108
   Gerbert's method, 123
Dodgson, Charles L., 240
Double false position, 113, 117
Doubling, 3
Duplication of the cube. *See* Constructions
Dürer, Albrecht, 135

$e$, 187, 188
Education, Greek, 45
Egyptian mathematics, 1–9
Einstein, Albert, 226, 237
Eleatics, 35–37

*Elements. See* Euclid's *Elements*
Equality sign, 134
Equations
   approximate solution, 104, 117
   classification, 15, 112, 115
   cubic, 115, 136–38, 183, 189–93
   linear, 8, 9, 101, 113
   quadratic, 12–13, 16, 61–62, 106, 112, 138
   quartic, 137
   quintic, 189–93, 222, 231
   symbolism, 83, 131–35, 140
Eratosthenes, 66
Eratosthenes' sieve, 66
Euclid, 54–55
Euclidean algorithm, 64
Euclid's *Elements*, 26, 54–65, 112
   book I, 37, 40, 42, 55–60
   book II, 61–62
   book III, 42, 63
   book IV, 42, 63
   book V, 49, 63
   book VI, 64
   book VII, 33, 64
   book VIII, 45, 65
   book IX, 30, 65
   book X, 48, 65
   book XI, 65
   book XII, 65
   book XIII, 48, 65
   postulates, 40, 223–27
Eudoxus, 49–53, 65
Eudemus, 59
Euler, Leonhard, 187–89, 201, 218–19
Exhaustion, method of, 51
Exponents, 132

False position, Rule of, 9
Fermat, Pierre, 168–71
Fermat numbers, 169
Fermat's last theorem, 170, 214
Fermat's little theorem, 169
Ferrari, Ludovico, 137, 222
Ferro, Scipione del, 136
Fibonacci, 124–26
Fibonacci numbers, 124, 156
Figurate numbers, 30–31
Finger reckoning, 122, 155

Fiore, Antonio Maria, 136
Fluxions, 179
Four-color problem, 232
Fourier, Joseph, 216–21
Fourier series, 216–21, 236, 254
Fractions
　decimal, 140, 212–13
　sexagesimal, 11
　unit, 5
　word origin, 134
Francesca, Piero della, 135
Fréchet, Maurice, 244
Function, 185, 188, 218–21, 223, 234, 238
　conjugate, 191
　symmetric, 191
Function space, 245, 246–47

Galileo Galilei, 152
Galois, Evariste, 231
Garfield, James A., 240
Gauss, Carl Friedrich, 43, 206–15, 249–50
Geometry
　axioms, 35–37, 244
　descriptive, 194
　differential, 214–15
　Euclid's *Elements*, 54–65
　foundations, 35, 243–44
　higher-dimensional, 245
　metric space, 244–45
　non-Euclidean, 212, 223–27, 237
　one-sided surface, 215
　of position, 189, 232
　projective, 171–74, 215, 254
Gerard of Cremona, 123
Gerbert, 122
Germain, Sophie, 213–14
Gibbs, Josiah Willard, 229
Gödel, Kurt, 247
Goldbach, Christian, 188
Golden section, 44, 125
Gravitation, law of, 178
Gregory of St. Vincent, 176
Group, 191, 222, 231

Hadamard, Jacques, 235
Hamilton, William Rowan, 227–29

Hanseatic League, 130
Harriot, Thomas, 134
Hausdorff, Felix, 244
Heat conduction, 216
Heath, Thomas L., 65
Heiberg, J. L., 65
Hermite, Charles, 177
Heron, 76
Higher-dimensional geometry, 245
Hilbert, David, 243–44
Horner's method, 104
Huygens, Christiaan, 176
Hypatia, 85
Hipparchus, 75
Hippocrates, 41, 63

$i$, 188
Imaginary numbers. *See* Numbers, complex
Incommensurability, 31–35, 48
India, mathematics of, 17–20, 104–10
Indivisibles, 52–53, 68, 152, 159
Infinite, 35–36, 129, 152, 241–43
Infinite series, 176, 182, 188, 201, 216, 222
Interest, compound, 187
Intermediate value theorem, 238
Intersection, 232
Intrinsic coordinates, 214–15
Irrational numbers. *See* Incommensurability and Numbers, irrational
Isocrates, 45

Journals, mathematical, 222

Kepler, Johann, 150–51
Kepler's laws, 150
Khayyám, Omar, 115
Königsberg bridge problem, 189, 201
Kummer, Ernst, 246

Lagrange, Joseph-Louis, 189–93, 222, 231
Laplace, Pierre Simon de, 193
Laplacian, 193
Leibniz, Gottfried Wilhelm, 184–85
Least squares, 206

Least time principle, 168
L'Hospital, G.F.A. de, 185–87
L'Hospital's rule, 186–87
*Liber Abaci*, 124
*Lilavati*, 107
Limit, 50, 181, 185, 221, 233
Line, continuity of, 238–40
Lindemann, Ferdinand, 177, 211
Lobachevsky, Nikolai Ivanovich, 223–27
Logic, symbolic, 184, 232, 247
Logarithms, 129, 142–49, 157, 158
  base ten, 148–49, 158
  Napier's, 142–48, 157
Lunules, 41

Machin, John, 177
Maclaurin, Colin, 186
Magic squares, 100, 116
Magnetism, 214
Map coloring, 232
Mapmaking, 139
Mathematics, definitions of, 45, 232, 243
Mathematics, word origin, 45
Matrices, 101–2, 229–31, 246
Maurolico, Francesco, 139
Mayan number system, 110
Mercator, Gerard, 139
Mercator projection, 139
Mechanics, 53, 129, 179, 193, 227
Mersenne, Marin, 173
Metric space, 244–45
Minuend, word origin, 134
Minus sign, 132, 133
Minute, word origin, 134
Möbius, Augustus Ferdinand, 215
Möbius strip, 215
Modular arithmetic, 212–13, 246
Monge, Gaspard, 194
Motion, 129, 152, 179
  Newton's laws, 179
Multiplication
  abacus, 99
  chessboard method, 132
  cross multiplication, 133
  Egyptian, 3–5
  grating (gelosia) method, 107

Napier's rods, 141
Russian peasant method, 128
scratch method, 128
symbol, 133
tables, 12
Music, 28, 45, 65

Napier, John, 141–48
Napier's rods, 141
Neugebauer, Otto, 10
Newton, Isaac, 178–84
Nicomachus, 30, 76
*Nine Chapters on the Mathematical Art*, 101
Noether, Emmy, 246
Noncommutative algebra, 229–31
Non-Euclidean geometry, 212, 223–27, 237
Normal curve, 207
Nowhere differentiable continuous function, 234
Numbers. *See also* Numerals
  binary, 256–58
  complex, 138, 207–8, 228, 236
  Fermat, 169
  Fibonacci, 124, 156
  irrational, 31–35, 48, 238–40
  perfect, 29–30, 93, 170, 188, 201, 212, 253
  prime, 66, 170, 188, 212, 235
  real, 238–40
Numerals
  abacus, 114
  Arabic, 114, 122
  Attic Greek, 37
  Babylonian, 10–11
  Brahmi, 105
  Chinese, 102
  Egyptian, 2
  Indian, 105
  Ionic Greek, 38
  Mayan, 110
  rod, 103
Numerator, word origin, 134

Olmec number system, 111
One-to-one correspondence, 36, 241–42, 257

Order types, 242
Oresme, Nicole, 129, 152

Pacioli, Luca, 132
Painting, 135
Pappus, 83, 167, 172
Parallel postulate, 55, 223–27
Parameter, 140
Partial differential equations, 193
Pascal, Blaise, 171
Pascal's theorem, 173–74, 254
Pascal's theorem, dual of, 215, 254
Pascal's triangle, 104, 174
Peacock, George, 228
Percent, 134
Perfect numbers. *See* Numbers, perfect
Permutations, 190, 222, 231
Perspective, 135
Pi, 71, 89, 139–40, 176–77, 188, 211
Planudes, Maximus, 127
Plato, 27, 46–48, 77
Plus sign, 131, 133
Polygons. *See* Constructions
Polyhedra, regular, 46–48
Poncelet, Jean-Victor, 215
Prime Numbers. *See* Numbers, prime
Principle of duality, 215
Probability, 171, 174–75, 186–87, 207
Proclus, 59, 85
Projective geometry, 171–74, 215, 254
Ptolemy, 43, 77–81, 83, 90–91
Ptolemy's theorem, 80
Pythagoras, 28–30
Pythagorean theorem, 33–34, 56–57, 64
Pythagoreans, 29–37, 39, 44, 76

Quadrivium, 85
Quaternions, 229, 255
Quotient, word origin, 134

Ratios of magnitudes, 49–50, 63
Real numbers. *See* Numbers, real
Rechenmeister, 130
Recorde, Robert, 134
Regiomontanus, 131
Rhind papyrus. *See* Ahmes papyrus
Ricatti, Jacopo, 186
Riemann, Bernhard, 221, 225–27, 235–37

Riemann surfaces, 236
Riemann zeta function, 235
Riese, Adam, 133
*Rig-Veda*, 18
Right triangles, integral sided, 15
Ring, 245–47
Robert of Chester, 123
Roberval, Gilles Persone de, 176, 199
Rolle, Michael, 186
Root, square. *See* Square root
Root, word origin, 113
Rudolff, Christoph, 133
Ruffini, Paolo, 222

Saccheri, Girolamo, 186, 224
Sacrobosco, 127
Second, word origin, 134
Sets, 232, 241–42, 244–46
  intersection, 232
  order types, 242
  union, 232
Shanks, William, 177
*Siddhantas*, 104
Similar triangles, 64
Sine, 75, 79, 105, 115, 217
Socrates, 27
Solid geometry, 65
Spiral, 71, 89–90
Square root
  algorithm, 83–85
  divide and average method, 13
  geometric explanation, 83–85
  symbol, 134
  table, 133
Squaring the circle. *See* Constructions
Stevin, Simon, 140
Structure, 247–48
Subtraction
  borrowing and repaying plan, 124
  complementary plan, 107
  simple borrowing plan, 127
Subtrahend, word origin, 134
*Sulvasutras*, 18–20
Symbolic logic, *See* Logic, symbolic

Tally, word origin, 134
Tartaglia, Niccolo, 136

Tautochrone, 176
Taylor, Brook, 186
Thales, 26–28
Theatetus, 48–49, 65
Theon, 83
Theory, word origin, 30
Three- and four-line locus, 167–68
Torricelli, Evangelista, 176
Transformation of variables, 229–31
Trigonometry, 75, 79–81, 105, 115, 131

Union, 232
Universities, 123

Vallée-Poussin, Charles-Jean de la, 235
Vector analysis, 229

Vibrating string, 217–18
Viète, François, 139
Volume
  cone, 52
  hemisphere, 68
  wine barrels, 151
Von Neumann, John, 247

Wallis, John, 176
Wantzel, Pierre Louis, 210
Weierstrass, Karl, 222, 233–35
Wessel, Caspar, 207, 229
Widman, Johann, 133

Zeno, 35–37
Zero, 10, 106